CROWOOD METALWORKING GUIDES

SHEET METAL WORK

CROWOOD METALWORKING GUIDES

SHEET METAL WORK

Dr Marcus Bowman

THE CROWOOD PRESS

First published in 2014 by
The Crowood Press Ltd
Ramsbury, Marlborough
Wiltshire SN8 2HR

www.crowood.com

This impression 2018

© The Crowood Press Ltd 2014

All rights reserved. No part of this publication may be reproduced or transmitted in any form or by any means, electronic or mechanical, including photocopy, recording, or any information storage and retrieval system, without permission in writing from the publishers.

British Library Cataloguing-in-Publication Data
A catalogue record for this book is available from the British Library.

ISBN 978 1 84797 778 6

Disclaimer
Safety is of the utmost importance in every aspect of the workshop. The practical procedures and the tools and equipment used in engineering workshops are potentially dangerous. Tools should be used in strict accordance with the manufacturer's recommended procedures and current health and safety regulations. The author and publisher cannot accept responsibility for any accident or injury caused by following the advice given in this book.

Typeset by Servis Filmsetting Ltd, Stockport, Cheshire
Printed and bound in Malaysia by Times Offset (M) Sdn Bhd

Contents

	Acknowledgements	7
	Safety First	8
	Introduction	10
1	Materials	13
2	Drawing and Developing Stages	21
3	Measuring and Marking Out	29
4	Cutting Sheet Metal	37
5	Making Holes	51
6	Bending Sheet Metal	61
7	Rolling, Beading, Flanging and Wiring	75
8	Joining	83
9	Forming, Pressing and Drawing	101
10	Surface Finishing	109

Projects:

1	Fuel Tank for a Model Aircraft	118
2	Motorcycle Ammeter in a Tin Can	124
3	Fluidizing Tank	131
4	Spigot for a Workshop Dust Extractor	134
5	Car Exhaust Expansion Chamber	140
6	Panels for a Sack Barrow	143
7	Folding Steam Iron Shelf	148
	Further Information	157
	Index	158

Acknowledgements

The author would like to thank those individuals and companies who expended time and effort to contribute photographs, illustrations and information for this book:

Rachael Bowman (photographic assistant)
George Nutt, Dennis Nutt and Chris Visscher at RMS Engineering Ltd
Dennis Nutt (Kilkerran)
Fred Anderson
Dewar Anderson
Sheryn and Steve Clothier at www.corrugatedcreations.co.nz
Tinplate Girl at www.tinplategirl.com
Dave Parker at Buxton Model Works
William Hurt at www.ageofarmour.com
Leslie Chatfield
Brian Walbey
John Saunders at www.NYCCNC.com
Jonas Boni at www.quarz.ch
Gary Tucker
RMT-Gabro and the M J Allen Group of Companies
EDMA Outillage
TRUMPF Group
Warren Machine Tools Ltd
RIDGID Tool UK
Irwin Tools
JD Squared Inc.
Jack Sealey Ltd

But most of all, a very special thank you to Hazel and Rachael, my long-suffering personal support team, for making the writing task so much easier.

Safety First

Pause for a moment, before you rush into the workshop to mangle metal for your latest project. The risks in most workshops deserve some careful thought, but the risks attached to working with sheet metal are amongst the greatest posed by any material. There are two particular risks: one associated with sharp edges, and the other the risk to hearing.

THE RISKS ASSOCIATED WITH SHARP EDGES

During the French Revolution, the guillotine was an effective way of separating head from body, and its effectiveness was largely because sheet metal behaves like a very slightly blunt razor. Because the edge of a sheet is so thin, the pressure it exerts is very high. You might get a nasty bruise or even break a bone if you strike your hand with a hammer or a large flat piece of metal, but strike your hand with the edge of a sheet and the injuries are likely to be much more severe, as the pressure instantly parts flesh, rips tendons, and slices through bone.

The edges left after a cutting operation usually have a fine, almost invisible burr which acts like a sharp razor saw, and the consequences of casually brushing a finger along such an edge are gruesome indeed. Yes: the edges of sheet metal deserve the utmost respect. Whenever possible, avoid handling the edges of sheets, and at the very least, wear good, heavy-duty, protective gloves of leather or Kevlar, and do not allow your hand to slide along an edge.

When moving sheets, avoid the edges and use mechanical aids such as a magnetic clamp or a sling with a metal hook.

Protect your feet from falling sheets by wearing good-quality protective boots, with steel toecaps.

In the UK, the Health and Safety Executive provides good guidance in Engineering Sheet No. 16 in the article 'Preventing injuries from the manual handling of sharp edges in the engineering industry'. Even if you are not an industrial user, you should avail yourself of this advice.

THE RISKS TO HEARING AND SIGHT

Striking sheet metal produces a distinctive noise that incorporates a great many very high notes and a ringing sound. Repeatedly striking metal, as when shaping a panel for example, produces an astonishingly high level of energy, and endangers hearing. The best protection is found in good-quality ear protectors with at least 30dB attenuation. Invest in the best protection you can, because hearing, once lost, cannot be restored.

Wear good eye protection made to proper industrial standards at all times, even if you wear spectacles. Proper wrap-around ski-style industrial goggles with impact-resistant lens material and soft seals around the face allow the comfortable use of spectacles. Eye protection should be able to resist puncture and impact. It should be comfortable to wear, and when it gets scratched, you should replace it. Think of it as a cheap investment in the most sensitive biological devices imaginable. Eyes cannot be replaced, and good protection is cheaper than medical bills.

CREATING A SAFE ENVIRONMENT

Sheet metal tools and machinery deserve your respect and your full attention, as a slip in concentration can easily lead to injury. Folders and press brakes present a temptation to fingers, welders may burn, and plasma cutters simply vaporize their target, so risk assessment and maintaining a keen awareness of potential risks will serve you well.

Read the instructions supplied with machinery and tools, and seek advice where you lack experience. Most skilled workers will be delighted to guide you and make you aware of hazards.

Take particular care when other people are present in the workshop. You have a duty of care to others, especially if they are unfamiliar with the workshop or the equipment. Watch those sharp corners if you are moving sheets when others are present; there is something resembling magnetic attraction between edges and unsuspecting bodies.

Throughout this book there are pictures of cutters and machinery that do not always have guards visible. These pictures are for illustration, to help you to see and understand what is being said or shown. Nothing in this book should be taken as an indication that we suggest, recommend or endorse working practices that are potentially hazardous or unsafe.

The sheet metal workshop is a creative place, and there are few pleasures as great as creating an object that not only serves a useful purpose, but may well be a thing of considerable beauty too. If that can be achieved in a safe environment, so much the better.

Introduction

ABOUT THIS BOOK

The aim of this book is to explain the tools and techniques required to produce sheet metal parts that are fit for purpose, accurately made, and attractively finished. This is a practical book, designed to be enjoyed by those of us who delight in making things, and it includes a set of projects that illustrate many of the techniques and tools.

This is a book about accurate methods and predictable, repeatable results. Although sheet metal lends itself to artistry and the kind of free expression that can lead to beautiful one-off objects, that kind of approach is not to be found within these pages. Nor is there any explanation of techniques of car-body manufacture or repair. Instead, the book is about making accurate drawings, then using those drawings to produce accurate components from sheet metal and small-diameter rod. The contents range widely over the materials, machines, tools and techniques which might be found in workshops and small factories. There is much talk of developing accurate shapes and templates, bending allowances, methods of attachment, and attractive finishing techniques.

There is some reference to stretching and shrinking, but this book does not deal with the handwork techniques required for car-body panels or artworks. There are methods and examples of punching, shearing and deep drawing, but little is said about spinning, as the applications of this in the smaller workshop are more restricted.

The approach to technical drawing covers the fundamentals of drawing layout, and looks forwards by explaining the basics of computer-aided design and drawing because that is the modern way. There is no attempt, though, to teach the use of one specific package, as that deserves a book or two on its own. Those readers who wish to develop their abilities in drawing by hand, still a surprisingly useful skill on site, must seek to learn from examples elsewhere in one of the established classics.

Because the book deals with modern methods, laser, water jet and plasma cutting are mentioned as being everyday industrial processes. We are perhaps not quite at the stage where laser cutters for metal are readily available at low prices, but these techniques now have their place in sheet metal work, and being able to order laser-cut parts is a natural consequence of understanding computer-aided drawing.

Despite the entrails of Britain's imperial past, and the rather mixed situation in North America, the units used throughout this book, including sheet metal thicknesses, are metric, as are thread sizes and specifications. That is intended to make the book accessible and usable right around the world.

This is therefore a practical book, to be enjoyed by everyone from the model engineer to the light industrial user; to be taken to the workshop and to be used to produce the kind of work that is real, usable, and rather beautiful besides.

SOME HISTORY

It is difficult to imagine modern life without metal objects – in fact, civilization and the history of human progress are inextricably linked with the discovery and use of metals. From the first discovery of gold around 6000BC, and its use in the manufacture of jewellery, to much more modern discoveries such as Lawrencium in 1961, the practical uses of metals have changed the way we live.

The first metals – gold, silver, copper, lead, tin, mercury and iron – brought a range of applications that transformed human capabilities. Gold was decorative but soft, and although not strong enough to be used for structural applications (because if the Golden Gate bridge had been made of gold, it would not have been strong enough to support itself), its softness allowed it to be hammered into very thin foil, paving the way, later, for the use of sheet metals.

Although soft by comparison with iron, copper was one of the more useful metals, because it was soft enough to be formed into shape, yet hard enough to be used for tools and weapons. Copper remains an extremely useful metal, in all its forms, and is widely used in sheet, tube and bar form in applications such as sheet roofing, electrical cables and conductors. And where would plumbing have been, without copper tube?

Bronze is an alloy (mixture) of copper and tin, and the discovery of bronze was so significant that it has given its name to a lengthy period in human history and development. In Europe, developments during the period from 3200–600BC were based on the uses of copper and bronze. Bronze is harder than copper, and can be used to make tools and weapons that are more durable (hard-wearing) than similar items made using copper alone. A bronze chisel, for example, holds a sharper edge for longer, and deforms less when struck, than a copper chisel. These characteristics of durability and hardness enabled significant progress during the Bronze Age.

But this period also proved the value of creating alloys, which are essentially mixtures of metal elements in various proportions, to make other metals that have different characteristics than the pure metals used in the mix. In more modern times, that same process is used to mix copper with aluminium, both soft metals, to produce an alloy which is much harder and stronger than the original elements, while retaining some of the useful characteristics of both.

Iron is an important element, but the period between the beginning of the Iron Age, and the industrial production of the more important steel (an alloy of iron and carbon with small quantities of other elements) was a long drawn-out process. Although the smelting of iron ore to produce iron was taking place in India by 1800BC, and steel was being produced sometime between AD200 and AD300, the mass production of industrial quantities of steel had to wait until the invention of the Bessemer converter in 1855. Only then could sufficient quantities of steel be produced to power the Industrial Revolution.

From the earliest times, engineers have been concerned with the application of materials, and producing and using a range of materials suited to particular tasks. The efficient use of materials is important because of cost, and good design makes effective use of the properties of particular materials, seeking to produce objects that are effective for their purpose, able to be manufactured efficiently, and at the lowest cost. While some objects might initially have been hewn from lumps of metal by a blacksmith, good engineering has seen the production of similar objects that use a fraction of those materials, cleverly arranged to take best advantage of the properties of that material.

For example, in Victorian times a lamppost might have been cast in iron – decorative, but massively constructed, heavy and expensive to produce. Modern lampposts are lighter and may consist of iron, aluminium and/or copper. While the style may be different the cost is lower, yet both old and new perform the same function.

Metal has been available in sheet form since the earliest days of the use of gold, when sheet, foil and rod were used to make jewellery, a practice that continues today. Although the softer metals such as gold, silver and lead could be hammered into sheet and foil, or later, rolled, industrial rolling mills to produce sheet of consistent thickness in quantity were only introduced in Europe and then England in the late sixteenth century.

In the early twentieth century, the adoption of sheet metal bodies for mass-produced cars brought about developments in the engineering design and shaping of sheet metal not only for the bodywork but for ancillary items such as the chassis, brackets, radiators and trim. Making the car as light as possible reduced the cost while increasing the performance, and a major contribution to low weight was the clever use of thin sheet metal. Latterly, the body itself was designed to act as the chassis, by using monocoque construction made possible entirely because of sheet metal design techniques. The development of alloys designed to allow deep pressing without excessive stress enabled the production of body parts featuring complex curves and folds.

The mass production of aircraft brought similar developments, mainly in aluminium and its alloys, and together with the automobile industry, became a great driving force behind developments in the design and use of sheet metal in manufacture. The heating and ventilation industry adopted sheet metal for ducting, and became a major user of this form of metal, which could be made sufficiently rigid yet allowed the use of techniques for forming complex bends, transformations from square to round, and ease of assembly and on-site fixing.

Together, the automobile, aircraft and heating and ventilation industries became, and remain, major employers, contributing in a very significant way to the economies of the major industrial areas of the world.

Sheet metal roofing, in use worldwide but perhaps more popular in the USA than elsewhere, makes use of the simple characteristic whereby changing the shape of a sheet, in this case by adding corrugations or repetitive folds, adds to its strength, allowing it to span a considerable distance without appreciable deflection. The use of zinc coating to protect steel made sheets sufficiently durable to be able to resist corrosion outdoors, adding to the capability of roofing materials made from sheet. Currently, steel and aluminium roofing and guttering is available in convenient, large, lightweight panels with protective coatings and factory-injected insulation, making the assembly of large roofs a relatively straightforward job.

Laser cutting and CNC forming of components from sheets has brought an increase in the complexity of design but ease of assembly with lower cost. These new techniques sit alongside the traditional techniques and machines that our forebears would recognize, and together they allow the design and fabrication of sheet metal parts for a large range of jobs, whether for mass production in a factory, small batch production in a smaller industrial unit, or on a one-off basis in a garage or home workshop. A collection of hand tools and some small machines will allow sheet metal to be cut, folded, rolled and joined, making the production of specific parts quite straightforward. Finishing those parts using some of the basic processes of filing, grinding and polishing will result in completed workpieces of which both the light industrial factory and the home workshop user can be proud.

1 Materials

This book deals with sheet, tube and rod in a range of metals, and all manner of items made from those basic forms.

CATEGORIES OF SHEET METAL

Sheets of metal are available in various sizes and thicknesses, varying from extremely thin 0.025mm (0.001in) to 150mm (6in) thick, or more. Working methods at the extremes of the range are quite different, so sheets are grouped in the following categories:

Leaf: A thin sheet, often of precious metal, varying in thickness from a few atoms thick to 0.025mm (0.001in). The most common example might be gold leaf.

Foil: A thin sheet, thicker than leaf and thinner than sheet. Foil varies in thickness from 0.025mm (about 0.001in) to 0.15mm (0.060in). Foil is typically available in the precious metals – gold, silver, platinum and palladium – as well as other semi-precious metals such as copper and brass. Other, more commonly available metals such as aluminium are often used as foils, for example baking foil.

Sheet: A flat sheet, thicker than foil, and thinner than plate. Sheet varies in thickness from 0.15mm (0.06in) to 6mm (0.24in). In this book, most sheet will fall between 0.15mm and 3mm (0.125in) simply because anything over 3mm thick becomes difficult to shape, and usually needs heavier machinery and a different set of techniques. The ease with which a sheet can be shaped varies with the properties of the material, with steel being more difficult to work than copper, for example, so the material determines how feasible it is to work with sheet in the range 3–6mm (0.125–0.25in).

Plate: Anything thicker than sheet (over 6mm) is termed plate.

SHEET THICKNESS

Sheet thickness can be expressed in millimetres or inches, but it is often easier to use standard reference numbers from the standard wire gauge (SWG), Brown & Sharp (B&S), which is sometimes called American wire gauge (AWG), manufacturer's standard gauge (USA), Birmingham gauge (UK). The number of different 'standards' is confusing, and the gauges used for ferrous (steel) and non-ferrous (aluminium, brass, copper) sheets are different, so this book will use metric thicknesses and the sheet metal gauges and imperial (inch) equivalents as shown in Table 1.

The following should also be noted:

- Metric sheet is now available in standardized thicknesses, so the full range of possible SWG thicknesses cannot be obtained in metric sheet
- Not all thicknesses are available in all sizes of sheet, nor are all thicknesses available from all suppliers
- SWG differs from AWG
- Sheet thicknesses are nominal, and exact thicknesses depend on manufacturing processes and tolerances, and may vary from one supplier to another, or one batch to another
- Some thicknesses are only available in particular finishes: 0.7mm steel sheet, for example, is often only available as Zintec galvanized sheet

SHEET STEEL

Sheet steel is available in a wide range of thicknesses, and is produced by rolling thicker material to a final size by reducing its thickness. The rolling process may be carried out when the steel is hot (above the recrystallization temperature of steel), or cold (below the recrystallization temperature). Recrystallization affects the strength of the finished sheet, and cold reduced (CR) sheet is stronger than hot rolled (HR) sheet, as well as having a better surface finish. It requires less pressure and less power to hot reduce a sheet, however, and recrystallization produces a more uniform internal structure as well as making the metal more ductile.

Sheet steel is commonly available in cold reduced form (sometimes referred to as 'cold rolled') to specification EN 10130:1991 (formerly BS1449-1:1983/CR4), European steel specification DC01:1.0330: this steel is

OPPOSITE PAGE:
A replacement hub cap for a 1924 Bullnose Morris Oxford, made from stainless steel tube and sheet, embellished with a brass ring, and topped by an aluminium hemisphere.

Table 1: Sheet steel (ferrous) thicknesses

SWG number	Metric thickness (mm)	Available metric thickness (mm)	Imperial thickness (inches)
		6.0	
4	5.893		0.232
5	5.385		0.212
		5.0	
6	4.877		0.192
7	4.470		0.176
8	4.064		0.160
		4.0	
9	3.658		0.144
10	3.251		0.128
		3.0	
11	2.946		0.116
12	2.642		0.104
		2.5	
13	2.337		0.092
14	2.032		0.080
		2.0	
15	1.829		0.072
16	1.626		0.064
		1.6	
		1.5	
17	1.422		0.056
		1.25	
18	1.219		0.048
		1.2	
19	1.016		0.040
		1.0	
20	0.914		0.036
		0.9	
21	0.813		0.032
22	0.711		0.028
		0.7	
23	0.610		0.024
		0.6	
24	0.559		0.022
25	0.508		0.020
		0.5	
26	0.457		0.018
27	0.417		0.0164
		0.4	
28	0.376		0.0148
29	0.345		0.0136
30	0.315		0.0124

commonly referred to as DC01 or CR4. The characteristics of grades D01 to D06 are listed in Table 2.

Hot rolled sheet is available to specification EN 10051:1991 +A1:1997 (formerly BS1449 –S1.2 (1991) HR4-Dry/P&O) European steel specification.

Fig. 1-1 shows a modern replica of a traditional suit of armour which demonstrates creative use of mild steel sheet and wire, as well as brass detailing. Fig. 1-2 shows an unusual use for a very traditional sheet material: corrugated iron, which is still widely used as a roofing and cladding material in many parts of the world.

Table 2: Sheet-steel grades D01 to D06

Grade	Properties	Applications
DC01	Drawing quality	Stretching, bending, roll forming
DC03	Deep drawing quality	Deep drawing, demanding stretch forming
DC04	Non-ageing deep drawing	Demanding deep drawing and stretch forming
DC05	Non-ageing deep drawing	Demanding deep drawing and stretch forming
DC06	Low-carbon non-ageing special deep drawing grade	The most demanding deep drawing and stretch forming

Table 3: Hardness of aluminium

Strength code	Properties
F	As manufactured; no heat treatment
O	Annealed, soft
T4	Heat-treated; naturally aged; stable

Strength code	Properties
T5	Artificially aged
H12	Hardened: quarter hard
H14	Hardened: half hard
H16	Hardened: three-quarters hard
H18	Hard

Table 4: Aluminium grades

Grade	Characteristics	Properties
1050–'O'	Temper: 'O' soft Surface finish: good (can be highly reflective) Anodizing: good Corrosion resistance: very good Can be more easily cold formed but has less strength than 1050A Welding qualities: excellent Machinability: poor	Tensile strength 80N/mm^2 Yield strength 35N/mm^2 (approx.) Shear strength 50N/mm^2 Elongation 42% Vickers hardness 20
1050A–H14 Formerly S1B H4	Temper: H14 ½ hard Surface finish: very good Anodising: good Corrosion resistance: very good Can be cold formed. Ideal for bending or spinning Welding qualities: good Machinability: poor	Tensile strength 100-135N/mm^2 Yield strength 75N/mm^2 (approx.) Shear strength 70N/mm^2 Elongation % (A50) 4-8 Brinell hardness 35
3003–'O'	Temper: H14 ¼ hard Surface finish: good Anodizing: good Corrosion resistance: very good Can be cold formed and has more strength than 1050A Welding qualities: very good	Tensile strength 200–240N/mm^2 Yield strength 130N/mm^2 (approx.) Shear strength 125N/mm^2 Elongation % (A50) 4-8 Brinell hardness 60
3103–H14	Temper: H14 ½ hard Surface finish: good Anodizing: good Corrosion resistance: very good Can be cold formed and has more strength than 1050A Welding qualities: very good	Tensile strength 140N/mm^2 (minimum) Yield strength 110N/mm^2 (approx) Shear strength 90N/mm^2 Elongation 8% (minimum) Vickers hardness 46
5251–H14 Formerly NS4 H3	Temper: H14 ½ hard Surface finish: good Anodizing: good Corrosion resistance: very good Can be cold formed and has more strength than 1050A Welding qualities: very good	Tensile strength 200–240N/mm^2 Yield strength 130N/mm^2 (approx) Shear strength 125N/mm^2 Elongation % (A50) 4-8 Brinell hardness 60

Fig. 1-1: Mild steel sheet, wire chainmail and brass detailing on a replica suit of armour. William Hurt is the maker. WILLIAM HURT

Fig. 1-2: Nikau detail: a tree made from corrugated iron. The makers are Sheryn and Steve Clothier at Corrugated Creations. SHERYN AND STEVE CLOTHIER

ALUMINIUM

Available in its soft form, aluminium sheet is also commonly available in harder grades, with properties as shown in Table 3.

Aluminium is readily available in sheets with the following specifications (see Table 4).

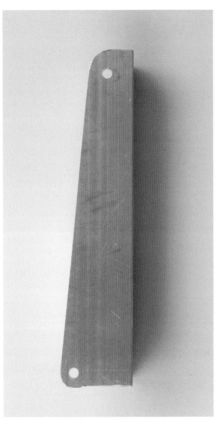

Fig. 1-3 shows a simple bracket made from commonly available aluminium angle.

BRASS

Brass sheet is available in different grades (*see* Table 5), and the colour is slightly different depending on the exact composition of each grade, varying from a yellow to a deeper gold colour. Not all thicknesses and sizes of sheet are available in all grades.

Brass is often plated with other materials, to protect its surface, or for decorative effect (*see* Fig. 1-4).

Table 5: Brass grades

Grade	Characteristics	Properties
CW505 Formerly CZ106	Machinability: good Formability: poor	70/30 'cartridge' brass with a high copper content
CW508 Formerly CZ108	Strength: excellent Hardness: excellent	High purity brass alloy
CW612 Formerly CZ120	Cold working: good Machinability: good	Leaded brass alloy Suitable for engraving
CW712 Formerly CZ112	Corrosion resistance: suitable for marine environments Strength: higher than normal Machinability: not as good as CW612	'Naval' brass, containing tin

Fig. 1-3: Bracket made from aluminium angle.

Fig. 1-4: Brass plates used as instrument housings; these items would originally have been plated with nickel or chrome. The vehicle in this photograph is owned by Dennis Nutt.

STAINLESS STEEL

The addition of at least 10.5 per cent chromium gives stainless steels improved corrosion resistance as compared to 'ordinary' steel. There are three commonly used grades of sheet: 304, 316 and 410, as listed in Table 6.

Stainless steel can be given an attractive polished finish, and is widely used in structural and decorative applications, despite carrying a higher price than mild steel. Fig. 1-5 demonstrates the difference between polished and unpolished finishes on stainless steel items.

COPPER

Most commonly available to specification ASTM B370, copper sheet is available in six tempers: 060 (soft), H00 (cold rolled), H01 (cold rolled, high yield), H02 (½ hard), H03 (¾ hard), and H04 (hard). Copper sheet is normally 99.9 per cent pure, but lead-coated copper is also available.

Copper is a soft metal and is very malleable, so it is easily beaten, pressed or rolled into shape. Cold-rolled copper is less malleable but stronger than soft copper, so that H00 ⅛ hard is less malleable but stronger than 060 (soft). Grades up to H04 (hard) are increasingly strong but have decreasing malleability.

Copper has excellent electrical conductivity, and good resistance to corrosion.

It can be easily welded, brazed or soldered. Fig. 1-6 shows a piece of copper sheet, as well as copper used as a hammer head, a task for which it is well suited despite its initial softness, as it work hardens with repeated impact.

Fig. 1-7 demonstrates the change in colour as copper cladding on a roof is exposed to the weather.

Table 6: Common grades of stainless steel

Grade	Characteristics	Properties
304	Corrosion resistance: good Formability: excellent Weldability: excellent Grade 304L recommended for welding	18% chrome, 8% nickel
316	Corrosion resistance: better than grade 304 Formability: excellent Weldability: excellent Grade 306L recommended for welding	Has added molybdenum. Better characteristics at high temperatures than grade 304
410	Corrosion resistance: moderate Formability: reasonable Weldability: poor Grade 306L recommended for welding	Can be heat-treated to produce excellent hardness properties

Fig. 1-5: Stainless steel discs, before and after polishing.

Fig. 1-6: Copper-headed hammer on a copper sheet.

18 • Introduction

Fig. 1-7: Copper changes colour dramatically when exposed to the weather.

Tinplate

Tinplate is thin steel sheet which has been coated with tin. Plating was formerly applied by dipping the steel sheet in a bath of molten tin. Current practice is to electroplate the sheet to apply a very thin coating of tin, typically 800 millionths of a millimetre (30 millionths of an inch) thick.

Tinplate has the mechanical properties of steel, but the non-toxic properties of tin, and is easily soldered. It is often used to make tins and similar containers (*see* Fig. 1-8), and is also used for manufacturing household items, and some electrical components such as shielding for electronic equipment.

Fig 1-8: Still commonly used for everyday items, tinplate is also available in sheet form.

NICKEL SILVER

Nickel silver contains no silver! It is an alloy of copper, nickel and zinc, often in the following proportions: copper 60 per cent, nickel 20 per cent, zinc 20 per cent. It is available in several forms, including sheet, and it has good electrical conductivity, good machinability, and its corrosion resistance is good, although it is not suitable for applications involving prolonged contact with acidic foods or drinks. It also solders easily. And it looks a lot like silver…

WIRE

Wire is available in a range of materials: steel, stainless steel, copper, silver, gold, aluminium and many others. In the same way that sheet is thinner than plate, wire is considered to be thinner than rod, and is usually specified by diameter, with the SWG numbers being widely used.

The term 'wire' refers to a single strand, although the term has come to be used loosely, with multi-stranded electrical conductors being called 'wire', for example. Several strands of wire, intertwined, is properly termed a 'rope', no matter how small its diameter.

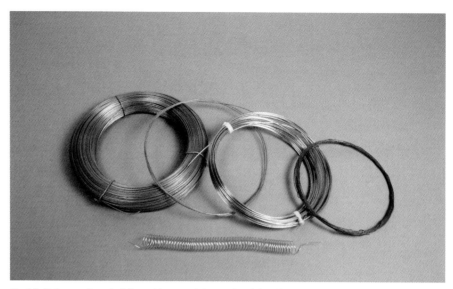

Fig. 1-9: At the rear, from the left: steel, brass, stainless steel and phosphor-bronze wire. At the front: nichrome wire.

Piano wire is made from spring steel which is hardened and tempered high-carbon steel. Its original use is for piano strings, but it is in common use for many other tasks that require the characteristics of a spring. Heating piano wire alters the temper and removes much of the spring, which distinguishes it from steel wire.

Nichrome wire is an alloy of nickel and chromium which may also contain iron. It has a high electrical resistance and is often used as a heating element in electric fires, electronic cigarettes, hair dryers and many other domestic products. It is also used in foam cutters.

Fig. 1-9 shows a small range of wire made from different materials.

2 Drawing and Developing Shapes

An artist might use a drawing to represent an object, and the style of the drawing might evoke a mood or highlight some aspect of the object to draw attention to its shape, its colour or its surroundings – but the way the drawing is interpreted by different viewers will depend partly on the viewer and their emotional connection with the subject of the drawing.

An engineer, on the other hand, uses a precise system of drawing which has strict rules for the use of lines and points so that everyone viewing the drawing will interpret it in the same way. An engineering drawing will provide all the information needed for an engineer to be able to make a particular object, and all the objects produced by following the drawing will be identical. Aside from any notes written in a specific language, engineers of any nationality should be able to interpret the drawing because it is based on a tightly specified set of visual rules.

In the UK, the rules governing engineering drawing are contained within the British standard 'Technical product documentation and specification' BS 8888:2011 (published in December 2011). BS 8888:2011 incorporates the International Standards Organisation standard ISO/TC 213 (GPS – Geometrical Product Specification) which applies across Europe and much of the rest of the world.

In the USA, the most widely accepted standards are based on the American Society of Mechanical Engineers standard Y14.5M (revised in 2009) and the American National Standards Institute standard Y14, both of which also recognize ISO/TC 123.

In this book, drawings are shown using third angle projection, where convenient, so that they are compatible with most systems and conventions. Furthermore all dimensions ('sizes') are shown in millimetres, because that is the preferred unit of measurement in the BS and ISO standards, and *may* also be used under ASME Y14.5M.

Although it is useful to have an understanding of both manual 'paper and pencil' methods and computer-based drawing techniques, this chapter provides only an overview of computer-based methods, and no practical details of paper and pencil methods. There are several classic manuals available for paper and pencil methods (*see* Further Resources), with many practical examples, and each of the major software packages has its own manuals and training materials. Both approaches are valuable, but deserve a much more complete explanation in greater depth than can be given here.

The rest of this chapter discusses drawing views, then illustrates the basic methods used to draw sheet metal parts using a computer-aided design package.

BASIC DRAWING VIEWS

Although sheets of metal begin life as flat surfaces in two dimensions with a uniform thickness, most assembled real-life objects are three-dimensional, so it is natural to begin with the completed object, and then work out the shapes and sizes of the component parts.

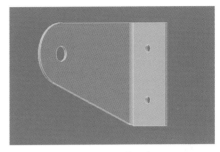

Fig. 2-1: Sketch of a sheet metal bracket.

Fig. 2-1 shows a sketch of a bracket. To show the full details of the bracket we need to be able to see it from different angles. Convention and the rules of engineering drawing suggest we should show the front, the side, and the view from the top. Here's how it is done:

- Imagine standing directly in front of the object, looking directly at it. Draw what you see, and put the drawing on the centreline of the page, a little less than halfway up the page (*see* Fig. 2-2).

Fig. 2-2: Drawing of the bracket, showing the front elevation.

OPPOSITE PAGE:
Vent stacks: stainless steel sheet and mesh.

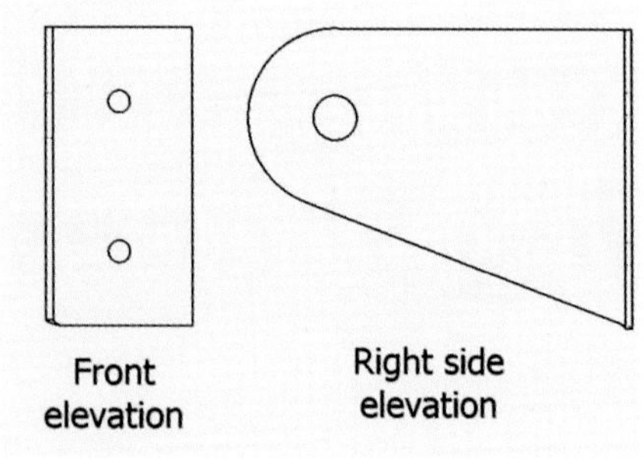

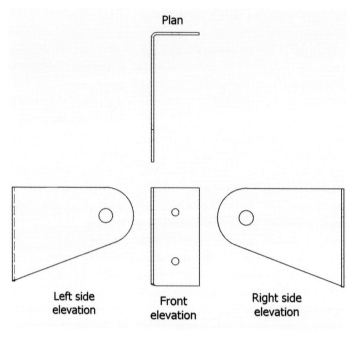

Fig. 2-3: Drawing showing the front and right side elevations.

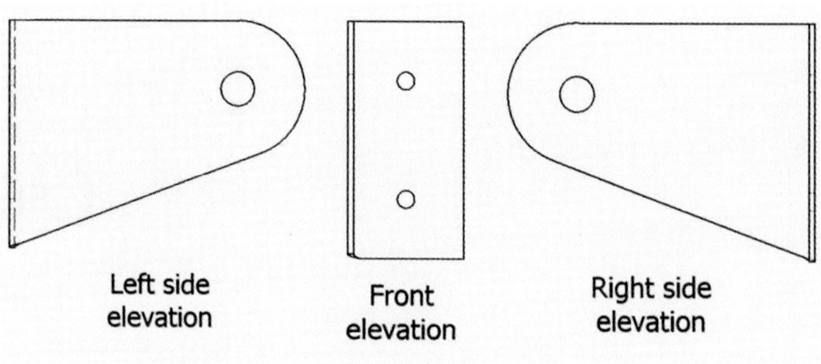

Fig. 2-4: Drawing showing the front, right and left side elevations.

Fig. 2-5: Drawing showing the front, right and left side elevations and plan.

This is the front elevation (elevation meaning simply 'view').

- Imagine standing to the right of the object, looking directly at the right side, then draw what you see. Put the drawing on the right of the page, on the same level as the front elevation (so the bottom of the bracket will be at the same level in both views): this is the right side elevation (see Fig. 2-3).
- Repeat for the left side of the object, putting the drawing on the left (Fig. 2-4). That is the left side elevation.
- Then imagine looking directly down from above; draw what you see, then put the drawing directly above the front elevation. That is the plan view.

The terms 'front' and 'side' are relative, as is the term 'plan', but they are useful in referring to views on the drawing because these are all accurately related to one another.

Those views are usually enough for an engineer to be able to see all the details on the object. In fact, one elevation is usually enough, in which case it is just called the side elevation. If the other side elevation shows hidden detail, include that view. There are occasions where additional views will be needed to see some details, but this basic set of three views is the best place to start, and is well understood by all engineers.

There are different ways of placing the views of the faces of an object in a drawing, and in order that an engineer should know which layout is being used, the symbol shown in Fig. 2-6 is used to denote third angle projection, which is a particular arrangement of views. If the symbol is as shown in Fig. 2-7 it denotes a different layout system, termed first angle projection, once common in some parts of the world but now superseded.

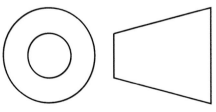

Fig. 2-6: Symbol denoting Third Angle projection layout.

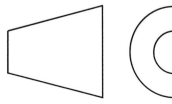

Fig. 2-7: Symbol denoting First Angle projection layout.

To allow accurate manufacture, dimensions can be added to a drawing, and lines used to connect the dimensions with particular features of the drawing. Place the dimension values above, on or below the dimension lines, oriented so that they are the right way up when the drawing is read from the bottom, or from the right-hand side.

COMPUTER-BASED DRAWING

In the beginning, some forty years ago, the approach to producing technical drawings using a computer was based firmly on a simulation of manual methods. In more recent times, drawing has become a subset of design, and computer-aided design (CAD) software now allows the user to draw in three dimensions, then produce the classic two-dimensional flat drawings often used for manufacture.

Drawing in three dimensions (3D) has many advantages, although the approach differs considerably from that of classic manual drawing. Packages such as Autodesk Inventor, Solidworks, RhinoCAD and Turbo-CAD 3D create 3D objects by using a set of tools to manipulate a basic 2D sketch to produce the 3D shape. This means that the kind of inner mental conceptualization required for manually drawing a 2D representation of the object – which very often does not exist at that stage – is, to a large extent, bypassed, and the user is very quickly working with a 3D object which can be rotated, scaled and altered on screen using the tools provided by the software.

Thus a physical drawing board with parallel motion mechanism and precision drafting head is replaced by screen, cursor, mouse and software tools, and once an object exists within the computer, the production of standard 2D drawings is a trivial exercise taking a few clicks of the mouse. Production of the developed shapes of component parts is a similarly simple procedure, taking little mental effort on the part of the operator.

But best of all, and most valuable of all, is the recognition within the software that an allowance must be made for the physical changes such as stretching and shrinking which take place when a sheet metal component is bent into shape. The software can calculate the effect of this, and take it into account when producing the development.

When drawings are produced manually using pencil and paper methods, a human draughtsman can examine a drawing and, armed with a knowledge of manufacturing methods, identify where sheet or tube will stretch or shrink during manufacture. They can then perform the calculations required to take account of this in the developed shape, and can redraw that development to compensate. But small changes to the design then mean re-examination and re-calculation so that, in effect, the human draughtsman must labour over something which to CAD software is a trivial task which can be performed automatically, accurately, and at high speed.

For this alone the software is worth its cost, and it is something that manual drawing methods alone cannot do.

Part of the cost of a software package is in the time required to master this new skill and this new way of thinking, and that should not be underestimated. Consistent continued practice is required to become accomplished in the use of any CAD package, and dedicated textbooks, online learning videos and user forums are all necessary elements of the learning task. Once mastered, however, design, the subsequent development of the shapes of parts, and the production of working drawings all become swift and painless, and that acts as a considerable aid to production of the parts.

One challenge, though, is that drawings are limited in size to the capabilities of the printer, and in order to have full-sized working drawings the completed files may need to be transferred to another computer, or to a dedicated print shop which possesses a large enough printer or plotter. Nevertheless many drawings are useful at a reduced scale, because on a CAD drawing, dimensions are always stated at full size, even though the drawing is scaled down to fit on to smaller-size paper, and it is relatively easy to transfer measurements to a sheet of metal.

Some projects, however, do demand full size printouts: for example, the construction of large naval vessels is based on printed drawings which are reproduced at full size for use as templates. With a hull size of, say, 154m, for recent naval destroyers, this is no small undertaking, even when the ship is built in sections.

SOFTWARE TOOLS AND METHODS

Software tools are based around sketches in 2D or 3D which form the basis for faces. Collections of faces form parts; collections of parts form assemblies; and parts and/or assemblies allow the creation of drawings.

Creating an Initial 2D Sketch

The purpose of this stage is to create a basic building block for a 3D model. The initial sketch need not be detailed or precise as it can be altered later, and alterations in the basic shape will make the subsequent 3D model update its shape automatically. An initial sketch is created using lines, circles, arcs, rectangles and similar shapes, using straightforward 2D drafting techniques. Shapes can be drawn at approximate size and in almost any location, with precise sizes and positions being specified later. Shapes can also be drawn precisely, by specifying coordinates for the vertices (corners).

Fig. 2-8 shows an initial 2D sketch consisting of a rectangle that will form the basis of a vice jaw protector (clam).

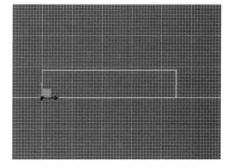

Fig. 2-8: Sketch of a rectangle that will form the basis of a sheet metal part.

Creating the Third Dimension

The 2D shape can be given a thickness to create a 3D object. Fig. 2-9 shows the rectangle thickened to form a sheet metal face (that is, a sheet component or part) using the face tool.

Fig. 2-9: Sketch thickened to form a sheet metal part.

Creating Features

A selection of tools can be used to add features to a part. For example, the flange tool adds straight-sided flanges, and allows the angle, height and bend radius to be specified. Fig. 2-10 shows two flanges added to a face using the flange tool, and Fig. 2-11 shows the flange tool dialogue box, which allows the height of the flange and the position of the bend to be specified.

Fig. 2-12 shows a third flange, and Fig. 2-13 shows a fourth flange, placed at an angle.

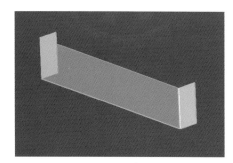

Fig. 2-10: Two flanges are added.

Fig. 2-11: Flange tool dialogue box.

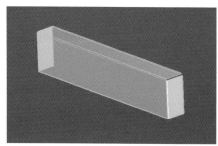

Fig. 2-12: A third flange is added.

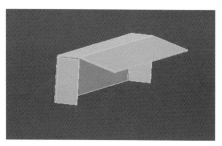

Fig. 2-13: A fourth flange is added at a 15-degree angle.

Creating a Development

A 3D object can be unfolded to create a developed shape, including bending allowances. The development can show dimensions, but it can also be altered, and changes will be reflected automatically in the 3D object and the underlying 2D shape. Fig. 2-14 shows the developed surface of the shape, created using the flat pattern command.

Fig. 2-14: The flat pattern development of the object.

Creating a 2D Drawing

A standard 2D drawing can be created using conventional layouts and projection systems (for example, third angle or first angle) or some other preferred layout, and additional auxiliary views can be added by clicking and dragging from any of the existing views, to automatically create and show the new view. Fig. 2-15 shows a basic drawing of the part.

Fig. 2-16 shows added dimensions produced not by human calculation, but by using the dimension tool to click on points on the drawing, which are then

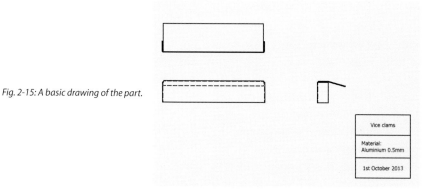

Fig. 2-15: A basic drawing of the part.

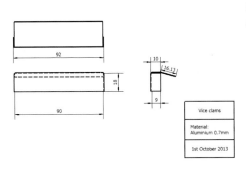

Fig. 2-16: Dimensions added using the dimension tool.

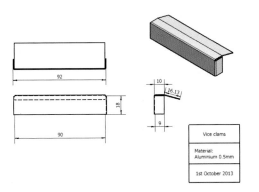

Fig. 2-17: A projected view is added to aid understanding.

dimensioned automatically by the software. Fig. 2-17 adds a projected view produced by using the projection tool.

Modifying the Part

Once drawn, a part can be modified at any stage, and Fig. 2-18 shows the part modified by adding rounded corners to the last flange, using the corner round tool. Modifications to the part cause automatic updating of all the other views of that part, so returning to the existing drawing shows that it has been changed to reflect the rounded corners (Fig. 2-19).

Printing a 2D Drawing

Having created a drawing, it can be saved as a PDF or image file, sent to a printer or plotter, or emailed to a print bureau.

MANUFACTURING THE PART

Once a part exists in its final form, there are several options for exporting the files and sending them to a laser cutter, a CNC punching machine, or a CNC folding machine. The exact requirements for the format of files may vary from one company to another, but the basic principle is that every drawing package can export in a transportable and universally recognized file format such as DXF (drawing exchange format), and every CNC manufacturing system should be able to import files having that format. Many manufacturing systems can also take 3D models in a transportable format or in the format native to any of the major CAD packages.

So, once the part has been created in software, it can be sent straight to a manufacturer for some operations or for complete fabrication.

Creating an Assembly

Once several parts have been created, they may be brought together in an assembly. This allows the designer to check the fit and the interaction between parts designed to work together. Any changes made at this stage will automatically update individual components. Fig. 2-20 shows an assembly

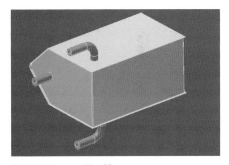

Fig. 2-20: An assembly of five parts.

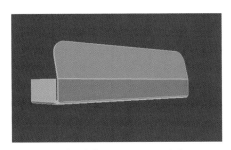

Fig. 2-18: The fourth flange is given rounded corners.

Fig. 2-19: The drawing is automatically updated to show the changes.

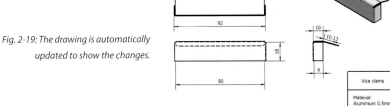

consisting of five parts: a U-shaped side/rear/side part, a top/front/bottom part, and three individual pipes, all placed accurately in relation to one another. The pipes continue inside the assembly.

Detail Decisions

One of the characteristics of software is that every feature of a drawing or a part must be specified precisely. To make life easy, a large number of default settings are incorporated, but any of these can be changed. Take corner details, for example. One technique is simply to assume that all corners meet neatly and that folding will not cause any difficulty, whether there are two or three or more surfaces around a fold or a corner. In the real world, things are not so simple, and it all depends on where faces and edges end up after a bend.

First, any bend has a specific radius, and no bend in the real world is absolutely sharp. The implications are tedious. Fig. 2-21, for example, shows that a bend will throw the adjacent face outwards (or in this case, upwards) by an amount that depends on the radius of the bend. The minimum radius will depend on the thickness of the material, and the actual displacement depends not only on the radius, but on the tendency of the material to stretch.

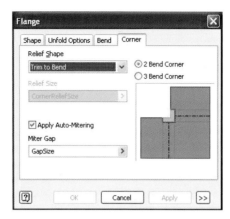

Fig. 2-22: Trim-to-bend corner relief.

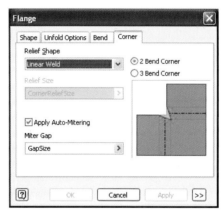

Fig. 2-23: Linear weld corner relief.

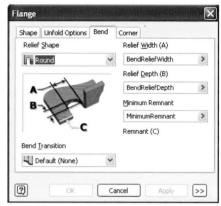

Fig. 2-24: Round relief shape.

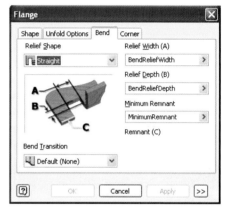

Fig. 2-25: Straight relief shape.

Fig. 2-21: The bend throws the top surface upwards by the radius of the bend.

There is also the matter of where exactly the bend begins. In Fig. 2-19 it is important that the folds result in a flat distance of 90mm on the inside of the part before any bend begins. The minimum radius of the bend is 0.5mm, so there will be a distance of at least 91mm between the insides of the two flanges, and that extra 1mm is a result of real physical constraints. A more generous radius would throw the flanges further out, while a smaller radius would draw them closer together.

What happens at the corners is that in many cases the folds will begin to interfere with one another, and there is a danger of the material tearing. Instead of simply hoping for the best, the software allows the corner reliefs to be precisely specified using a series of settings in a menu. Figs 2-22 and 2-23 show two of a wider range of options for corner reliefs, and Figs 2-24 and 2-25 show two of a range of possible bend relief shapes.

Fig. 2-26 shows a circular bend relief

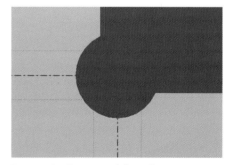

Fig. 2-26: Circular corner relief, on a flat pattern.

shape as it would appear on the flat pattern, easily produced by drilling or punching a hole centred on the junction between bend lines, and large enough to ensure that all of the curved surface of the bend can be formed without interfering with adjacent bends. Fig. 2-27 shows the result on the folded part.

Drawing and Developing Shapes • 27

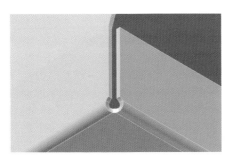

Fig. 2-27: Circular corner relief on a folded object.

Adding Features

Most drawing software incorporates a wider range of additional tools to make common tasks easy. Fig. 2-28 shows a 2D sketch of a circle, drawn on an existing part by using the circle tool. The part already has a hole in the centre, and a hole on the sketched circle.

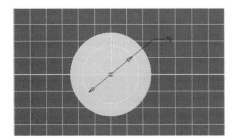

Fig. 2-28: Sketch drawn on an existing part. The sketch shows a circle with a centre mark indicating the position of a single hole to be used to make a pattern.

The part requires a pattern of six holes spaced equally around the circle, at the same radial distance from the centre. Fig. 2-29 shows the circular pattern tool dialogue box

Fig. 2-29: Circular pattern dialogue box.

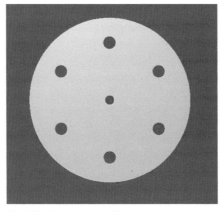

Fig. 2-30: Circular pattern of holes.

specifying this pattern, and Fig. 2-30 shows the result.

Solving Problems

The lofted flange tool creates a lofted sheet running between two different shapes, such as a circle at one end and a rectangle at the other. This is a powerful tool which can literally solve a common but complex drawing problem with a couple of clicks of the mouse.

Fig. 2-31 shows two rectangular shapes drawn in 3D, a specific distance apart. The lofted flange tool is used to automatically draw the sheet shape running between these rectangles to produce a transformer (Fig. 2-32). The shapes need not be restricted to rectangles, and transformers can be produced between circles and rectangles, or any other pairs of shapes.

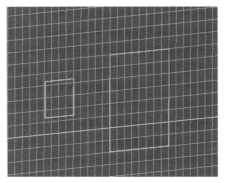

Fig. 2-31: A 3D sketch of two rectangles, a specific distance apart.

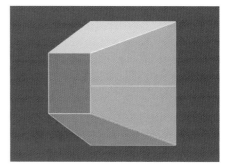

Fig. 2-32: A transformer produced using the lofted flange tool.

The rip tool can then be used to cut the shape (as shown by the line running along the side of the shape in Fig. 2-32), and the flat pattern command produces the flat developed shape of the transformer shown in Fig. 2-33, as well as the positions of the fold lines required to produce the object.

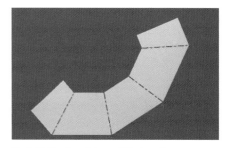

Fig. 2-33: Developed shape of the transformer, produced using the flat pattern tool.

SO MUCH MORE

The examples above are far from exhaustive, and the major software packages contain many more features, tools and commands to simplify drawings and the associated calculations. A full understanding of any particular package demands further study using dedicated books and study materials appropriate for that package.

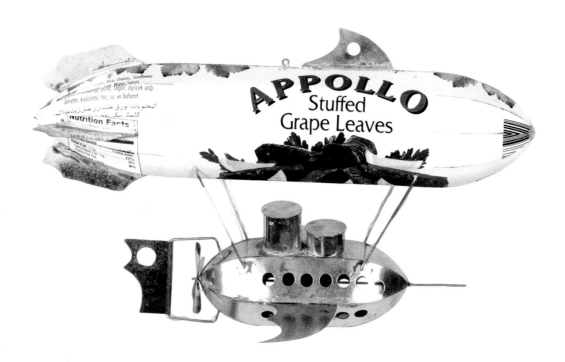

3   Measuring and Marking Out

Unless you are making a freehand sculpture, most sheet metal projects need to be marked out on to the sheet before cutting, folding, shrinking or stretching takes place. Cutting and folding require precise marks on the sheet to indicate where to cut or to begin a bend. Stretching and shrinking usually require an approximate indication of the area where the work is to take place.

Marking out for stretching or shrinking is often done using a large felt-tipped marking pen to indicate an area on the sheet. Great precision is not required, so we need say little more about those marks in this section. Marking out for cutting and bending, on the other hand, must be carried out precisely, carefully and accurately.

For some projects, marking out can be done by making a paper template, sticking that to the work, and using it as an indication of where to cut or bend. That technique is often recommended when using tinplate, or for model railway metalwork. The challenge is in printing a pattern precisely, and in avoiding distortion caused by the glue softening the paper pattern. With care and skill, this method can produce very usable results, but it is not suited to highly accurate work, or to a production run of parts.

A variation on this method is to stick plain paper to the metal, then draw the pattern on to the sheet before cutting or bending. That is then similar to the engineering solutions of marking out directly on to the sheet, but does allow the use of pencil or pen to draw the cut or bend lines. The disadvantage is that pencil and pen lines are rather thick and cannot be positioned quite as precisely as the engineering techniques of marking out allow. It is, though, a viable method, but, like making a drawing on paper then sticking it on to the work, it is not suited to a production run.

The starting point for most sheet metalwork projects is an accurate drawing. There are some elements of fitting where a partially finished workpiece is trimmed to fit an existing part, perhaps in adding a new duct to an existing installation where the existing parts are difficult to measure or have not been made to an expected shape, but most new work is designed and drawn before manufacture. Drawings can be created in a computer drawing package or can be drawn on paper using traditional techniques, but in either case the end result is a drawing which can then be used to guide the cutting and bending operations to be carried out on a sheet to produce the finished workpiece.

CAD, CAM AND CNC

Ideally, cutting and bending would take place without having to make marks on a sheet. That is possible using computer-aided design (CAD) to create drawings, computer-aided manufacturing software (CAM) to provide a set of instructions for machining based on the drawings, and computer-controlled machinery (CNC – where the NC stands for the old term 'numerical control') for cutting and bending.

CAD software is very useful because it

Fig. 3-1: Sheet component made by cutting out on a milling machine.

automatically calculates allowances for bending and can easily produce the 'development' or folded-flat shape of an object to give the dimensions needed for a shape. CNC sheet metal machinery is not readily available in the small workshop, with the possible exception of a CNC milling machine which can be used to cut a shape from sheet using a rotary milling cutter (Fig. 3-1).

Although there is always a challenge in designing a method of holding the sheet on the milling machine table, using this kind of machinery can allow a workpiece to be cut to size without marking out the sheet. This can help achieve accuracy because it avoids a stage that depends on human eye–hand coordination, and so reduces errors. What this method is not good at is producing square internal corners, because each milling cutter has a radius (see Fig. 3-2). This effect can be eliminated by altering the design of a sheet metal part, to avoid the need for sharp internal corners (Fig. 3-3), in many cases.

OPPOSITE PAGE:

Tinplate airships. TINPLATE GIRL

30 • Measuring and Marking Out

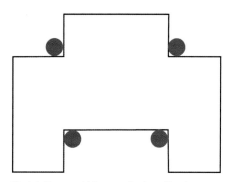

Fig. 3-2: Corners which cannot be sharp if cut by an end mill.

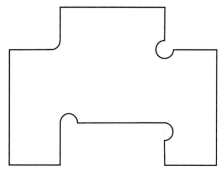

Fig. 3-3: Four different ways of redesigning corners to be cut by an end mill.

CNC laser cutters, plasma or water jet cutters can produce finely detailed parts, accurately cut, without marking out, and this is useful for small and medium-sized parts, but not for one-offs or for small runs of components, as initial set-up costs can be relatively high: this means that many parts can be produced for not much more than one or two.

MARKING-OUT TOOLS

For single items or small batches, or for work carried out on site, most sheet metal projects require marks to be made on a sheet to indicate the positions of cut lines, bend lines, hole centres and similar features. These marks can be produced by hand, from positions indicated on a drawing. Whether the drawing has been produced by CAD or by hand is unimportant. However, many sheet metal workpieces are too large for a CAD system to produce life-size 1:1 scale printouts, so we are often working from a scale drawing. Some drawings are produced on site, by hand, and may be created by drawing on a convenient floor, simply because of their size. Dimensions can then be taken from that drawing and transferred to a sheet.

Once cut, the workpiece can easily be compared to the life-size drawing. Scaled-down drawings must have their principal dimensions transferred to the metal sheet. If the drawing is full size in the CAD system, the dimensions printed on the drawing will be the real sizes. Scaling down the printout does not change the dimensions held in the CAD system or printed on the drawing, it is simply that the drawing itself is printed smaller than life size. Work to the dimensions on the drawing, but do not measure any sizes directly from the drawing because any errors in measurement will be magnified by the scale of the drawing.

Marks made directly on to a sheet of metal may be visible in good light, especially if the marking tools are sharp. As an aid to visibility, the surface of the sheet may be coated with a contrasting colour using chemicals or fluids.

Traditional marking-out fluid is blue and is spread on the sheet using a paintbrush, but spray versions are available and may be more suitable for large areas. Once the fluid is dry, scribed lines show up clearly because their metallic colour contrasts with the blue (*see* Fig. 3-4). Some metalworkers use felt-tipped pens with a broad tip, designed for posters and presentations. These pens leave a 12–14mm broad line and are suitable for smaller work or specific areas of a sheet, but would be laborious to use to cover a large area. Note that layout blue is not the same as engineers' blue, which is used for a different purpose and is, in any case, non-setting.

Measuring and Making Marks

Measurements can be made with a ruler, and marks made by a scriber with a hardened steel or tungsten point (Fig. 3-5). A good ruler has fine graduation marks and a non-reflective surface to make it easy to read. It is then possible to work to graduations showing 0.5mm, 1/32 or perhaps 1/64in, depending on the accuracy required by the job. Myth and legend assert that some skilled workers can work to an accuracy of 0.2mm (1/128in), but this is doubtful for ordinary mortals, and is difficult to sustain during cutting and bending operations when working to a mark. Some people find optical aids such as a magnifying glass or a loupe to be useful, and most people find a good light essential.

A straight-edge is like a ruler, but does not

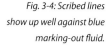

Fig. 3-4: Scribed lines show up well against blue marking-out fluid.

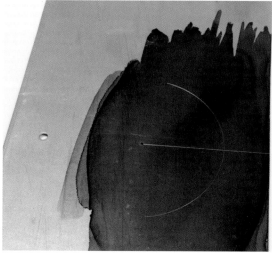

Measuring and Marking Out • 31

Fig. 3-5: Rulers and scribers.

Fig. 3-6: Engineer's squares.

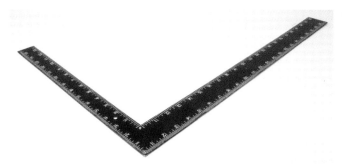

Fig. 3-7: Framing square.

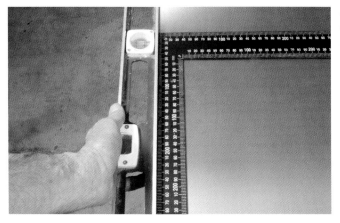

Fig. 3-8: Use a straight-edge to make sure a framing square is aligned with an edge.

have graduations. Its purpose is to provide a truly straight edge, and it is a precision instrument which should be treated with care. Usually, one edge is full thickness, while the other is bevelled. Either can be used as a reference. Some straight-edges look more like steel beams and are designed for use as a comparison when working with materials that are not in sheet form. Their edges can still be used on sheet, but they are not so convenient to use. Straight-edges may be supplied with a calibration certificate, but that implies an accuracy beyond that needed for sheet metal layout. Compared to a good ruler, a good straight-edge is an expensive item and not the kind of precision instrument you would use on most everyday jobs.

Squares

Use an engineer's square (Fig. 3-6), a layout square, a framing square or a roofer's square (Fig. 3-7) to ensure that a line is at right angles to an edge – and it helps to use as large a ruler or square as possible. A ruler can be placed in contact with the blade of an engineer's square, so that the square locates against the edge, with the blade setting the direction for the line, and the ruler providing the measurement.

Using the layout square, the edge of one arm is aligned parallel to the edge, and the other arm lies at right angles and provides a scale. The accuracy of the direction depends on maintaining the alignment of one arm with the edge of the sheet (or another appropriate feature on the workpiece). Initial alignment can be done by holding a straight-edge against the edge of the sheet, then butting the arm of the layout square against that edge (*see* Fig. 3-8).

A combination square (Fig. 3-9) carries an adjustable ruler and can be aligned with an edge in the same way as an engineer's square, but suffers from the disadvantage that on most models, the ruler zero cannot be aligned with the edge of the sheet.

Fig. 3-9: Combination square.

Fig. 3-11: Dividers.

Scribers

The purpose of a scriber is to make marks on a (usually) metal sheet, so the point of the scriber needs to be hard and sharp. Tungsten-pointed scribers retain their point over long use, while a steel scriber may need to be resharpened at regular intervals. Steel scribers may have a conical point, or the end may be forged into a broad knife-like blade (*see* Fig. 3-5).

The key skill with a scriber is in ensuring the point is as close as possible to the required position for the mark. When using a scriber against a ruler, tip the scriber so that the point lies against the ruler (*see* Fig. 3-10). A long slender point is of more use than a short stubby point, as the slender point will lie close to the ruler yet more upright as the scriber is tilted, while the stubby point needs to be tilted further. The more the tilt, the less effectively the scriber can make a mark.

A scriber blade can be run along the edge of a ruler quite easily, although it, too, must be tilted to allow the marking edge to lie as close to the ruler as possible, unless the scriber is sharpened so that one edge is straight and the other is bevelled. Using the straight-edge against the ruler allows the scriber to remain vertical while making its mark. One disadvantage of the blade scriber is that it cannot work into a corner or stop as precisely as a pointed scriber. The blade does, however, provide good guidance as it runs against the ruler.

Marking Out Circles and Arcs

Circles and arcs are marked out using dividers which, like scribers, have sharp points (*see* Fig. 3-11). Set the distance between the points to the radius of the circle, locate one point on the centre of a circle, then, keeping the dividers upright, turn them so that the other point marks the circumference of the circle.

Standard dividers have a limited reach, although larger dividers are available for larger drawings. The traditional design suffers from the disadvantage that the angle of the scribing point changes as the legs are spread, and with very wide openings the efficiency and accuracy of the scribing action is greatly reduced.

> ### COMPASSES AND DIVIDERS
>
> Compasses have one sharp point and the other point holds a pencil lead. They are used to draw circles on paper. Set the distance between the points to the size of the radius of the circle, press the sharp point lightly into the centre point of the circle, then keep the compasses upright as you turn them to draw the circumference of the circle. Keep the lead sharp by rubbing it at an angle on emery paper to give it one sloping face (Fig. 3-12).
>
> Dividers have two sharp points and are used to draw circles on sheet metal, plastic or other non-paper surfaces by scratching instead of drawing.

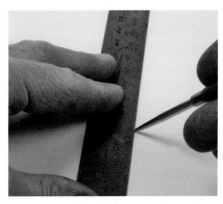

Fig. 3-10: Tip a scriber slightly to ensure the point lies close against the ruler.

Fig. 3-12: Sharpen a compass lead by giving it one sloping face.

Measuring and Marking Out • 33

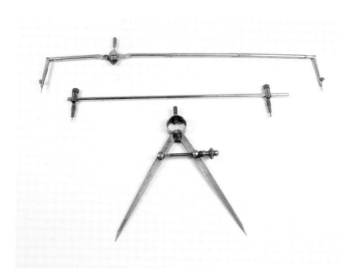

Fig. 3-13: To draw larger circles, trammels or beam compasses locate more securely than widely spread dividers.

Fig. 3-16: A magnet may be used to make a jenny divider.

An alternative is a beam compass (sometimes called a trammel) fitted with a scribing point instead of a pencil lead. The beam can hold the points vertically whatever the distance from the centre mark (see Fig. 3-13).

Where the circle is very large, as when drawing a large layout on a floor or a large sheet, an approximate circumference can be drawn using string as the compass. Hold one end at the centre mark, and hold a piece of chalk in a loop of the string at the radius of the circle, then move the chalk around in a circle by holding the string taut as it moves. An improvement would be to hold a pointer at the centre of the circle, and loop the string loosely around that pointer. Keep the pointer vertical as the string swings around while the circle is drawn. This is not a terribly accurate method, but it can be useful for providing an approximate indication. It all depends on the accuracy required.

Marking Out Lines

A line can be set out parallel to the edge of the sheet by measuring and marking, then joining the marks. For lines not too far distant from the edge, a pair of odd-leg dividers (sometimes called 'jenny' dividers) can be used. These have a lug on one leg which is held against the side of the sheet (see Fig. 3-14) and the other leg comes to

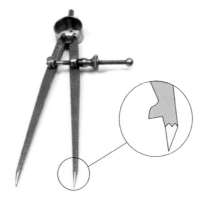

Fig. 3-14: Jenny dividers have an added spur on one leg.

Fig. 3-15: Adjustable parallel scribing guide, which relies on coordinating the movements of both hands.

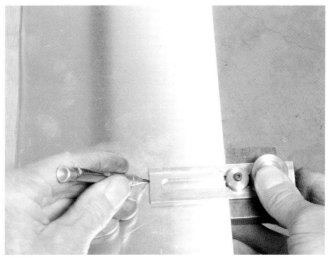

a point which acts as a scriber. Hold the dividers square to the edge, and run them down the sheet. The trick is maintaining the square orientation.

A variation on this is a device which locates against the edge but holds itself square, scribing a more accurately spaced line (see Fig. 3-15). With care and a gentle touch, an even simpler device will work for non-ferrous sheets and dividers with legs which have a flat inner face: just add a small but powerful magnet to one leg of a divider, a few millimetres above the point (see Fig. 3-16).

Setting out a line which is offset from another line or other feature on a sheet can

Fig. 3-17: Using a ruler to mark an offset line.

be done using a ruler. The most convenient way to do this is to use the ruler scale upside down so that the mark for the new line can be made against the zero end of the ruler (*see* Fig. 3-17) – but make sure the ruler is square to the work by aligning graduation marks on both edges of the ruler.

Setting out a line at right angles to an existing line can be done by measuring and making a mark where the new line crosses the existing line, then using the layout square to set the direction of the new line. This can also be done by using dividers, as shown in Fig. 3-18. From the point shown in red at which the new line is to cross the old, draw a circle to cross the existing line (points 1 and 2 on the orange circle). Taking a larger radius, and using 1 and 2 as centres, draw two identical arcs (shown in blue). Create the new line (shown in green) by joining the points at which the arcs cross (points 3 and 4).

Setting out a line parallel to an existing line can be done by using a ruler to measure and mark out a minimum of two points (or short line segments) from the existing line. Then join the new marks with a ruler or straight-edge. For longer lines, set out more marks, then scribe the new line by joining the points. Where the new line is defined by two marks, a third mark between the other two acts as a check on the straightness of the new line.

It is also possible to set out a new line by using points on the existing line as the centres of circles. Make the radius of the circles the same as the distance between the lines. Lightly centre punch the points, use dividers to draw the circles, then draw a new line tangent to both circles (*see* Fig. 3-19).

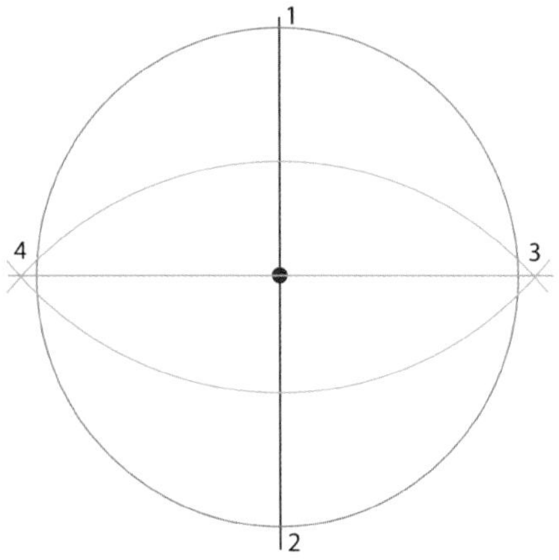

Setting Out Marks on a Circle

Centre Marks

A centre mark is made using a centre punch (Fig. 3-20), but it is best to use an optical centre punch (Fig. 3-21) to locate that point and position the punch accurately.

Fig. 3-20: Right to left: Pricker, standard 60-degree and heavy duty 90-degree point centre punches.

Fig. 3-21: Optical centre punch, including a punch with a finer point.

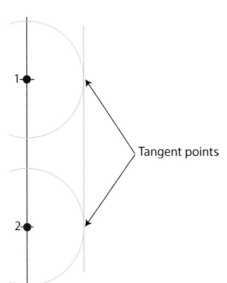

Fig. 3-18: Using dividers to set one line at right angles to another.

Fig. 3-19: A parallel line can be drawn as a tangent to two identical circles.

However; many models provide only a centre punch with a large included angle at the tip. This is not suitable for sheet, and should be replaced by an accurately made punch with a finely tapered point.

Whether using an optical aid or marking out unaided, use a pricker punch (*see* Fig. 3-20) before a centre punch. The pricker punch has a fine point, and a long taper to allow the point to be seen while the punch is vertical. If the point to be marked is at the intersection of two lines, drag the pricker gently across the sheet until its point drops into one of the scribed lines. Then run it along that line until it clicks against the intersecting line. Tap gently to make a fine mark. Locate the centre punch in that mark, and tap firmly. Keep your punches really sharp, for fine work.

Radial Lines or Hole Positions

Radial lines or hole positions can be set out on a circle using dividers by making marks on the circumference of the circle at regular intervals. The basic method is to set the dividers to a guesstimate of the distance between adjacent points, then step the dividers around the circle, from one point to the other, pivoting first on the end of one leg, then the other. If the last point coincides with the first, the dividers are set correctly. If not, adjust the setting and repeat the stepping-out. Keep adjusting and checking until the correct distance is obtained. Then repeat the stepping-out, scribing short arcs across the circumference at each position to indicate the ends of the radial lines or the hole centres (*see* Fig. 3-22). The basic method can be shortened for some numbers of lines or holes.

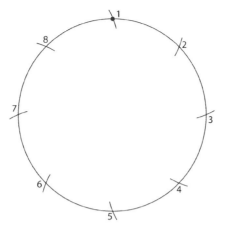

Fig. 3-22: Dividing the circumference of a circle equally.

For four holes, draw a diameter from one point, through the centre to the opposite point. Set the dividers to a distance larger than the radius. Locate the dividers on the first point and scribe a circle; repeat for the second point. Join the two points where those new circles intersect. You could simply draw the first diameter, then use a square to draw a second diameter at right angles to the first, but scribing the other two circles is often more accurate or more convenient.

For six holes, set the dividers to the radius of the circle, then step round the circumference.

For twelve holes, set out six equally spaced holes first. Then, without changing the setting of your dividers, swing an arc from each of two adjacent points. Draw a line between the intersection of the two arcs and the centre of the circle. Where that line cuts the original circle is midway between two of the original points. Use that as the starting point to step out a second set of six points. This results in twelve equally spaced points.

A Circle Tangent to Two Existing Lines

Sometimes, it is necessary to draw a circle tangent to two existing lines, perhaps to create a radius in a corner (*see* Fig. 3-23).

Set the dividers to a conveniently large distance. From the intersection of the two original lines (shown in black) at point 1 (the red dot), draw an arc (shown in blue) that cuts those original lines at points 2 and 3. From points 2 and 3 draw identical arcs (shown in brown) that intersect to give points 4 and 5, then draw a line (shown in green) through those points to the intersection of the original two lines (point 1). The centre of the tangent circle will lie on that green line. The further along the line, the larger the radius of the tangent circle. Pick a suitable point on the green line and draw a circle that just touches both the original lines.

To create a tangent circle with a specific radius, draw a line offset from one of the original lines by the radius of the tangent circle. Repeat for the second line. Where the two new lines cross is the centre of the tangent circle; that centre point should lie on the green line.

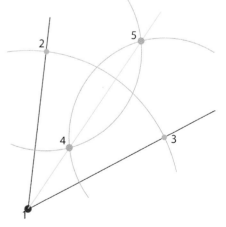

Fig. 3-23: Drawing a circle tangent to two lines.

4 Cutting Sheet Metal

STRAIGHT CUTS AND GENTLE CURVES

Straight or gentle curves can be cut in sheet using a variety of tools and methods. The choice of method may depend on the amount of distortion that can be tolerated, as well as the position of the cut within the sheet.

Snips and Shears

Snips and shears act in the same way as scissors, cutting sheet like cutting card, and the term 'shear' describes the cutting action of these tools. A shearing force acts parallel to the cross-section of a sheet (see Fig. 4-1) and does not involve bending. Shears and snips apply equal forces in opposite directions across a cross-section so that one side is pushed down and the other is pushed up at the same time. Shear force acts parallel to the cross-section, producing a shear stress, and when that stress exceeds the shear strength of the material, it fractures (shears) across the cross-section.

The cutting action of snips shears the sheet at the cut, but does not remove metal. Part of the process compresses the sheet along the edges of the cut, leaving stresses in the edges as well as tending to stretch the edges. On cuts around curves, the complex forces involved in shearing and bending can leave an edge that looks rather chewed (see Fig. 4-2), and this may not be acceptable on some work. Although this can be dressed afterwards, this means additional work, and the final position of the edge may differ from the planned line of cut.

Snips are available with straight blades or with substantial side cheeks (Fig. 4-3), and can cut both straight lines and gentle curves. Snips with curved blades (Fig. 4-4) assist with cutting tighter curves, but use the same scissor action. With these snips, in addition to the distortion caused by the scissor action shearing the sheet, the sides of the sheet will also tend to curl as the sheet parts around the jaws of the snips. This is especially true of snips with substantial side cheeks. These cheeks are

> ### STRESS
>
> Force is a push or a pull. It has a direction as well as a size, so it is a vector quantity. Typical units of force are Newtons (SI units, in the metric system) or pound-force (imperial system).
>
> Acting across a cross-section, force does not tell us very much. Which is more: a small force acting across a large cross-section, or a large force acting across a small cross-section? To decide, we need to know how much force is acting across the same cross-sectional area. That's a measure of stress.
>
> Stress is the force acting per unit area and is normally expressed as Newtons per square metre (known as Pascals, or Pa for short, in the SI metric system), or pounds per square inch (psi, in the imperial system).

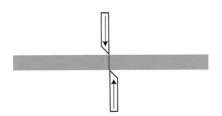

Cross section at red line is in shear
Left side is forced down
Right side is forced up

Fig. 4-1: Shearing action across a sheet.

OPPOSITE PAGE:
RMT-Gabro sheet metal notcher on a stand.
THE M J ALLEN GROUP OF COMPANIES

Fig. 4-2: A chewed edge as a result of the shearing action of snips.

Fig. 4-3: Snips with side cheeks.

Fig. 4-4: Snips with curved blades.

Fig. 4-5: Compound action aviation snips.

Fig. 4-6: Serrations on the jaws of aviation snips.

Fig. 4-7: Serrations on an edge left by aviation snips.

shaped so that the cut edges are guided over the cheeks, but on a longer cut this does mean the edges are widely separated as the snips work their way further into the sheet, leading to significant bending and distortion.

To minimize the distortion on one side of the cut, keep that jaw flat to the sheet and in contact with the sheet during the cut. Cutting a thin strip from one side of a sheet will result in a spiral coiled offcut, which will be distorted in more than one plane. Where the strip is narrow, it will tend to curl and twist, making it difficult to straighten it afterwards.

Compound Action Snips

Compound action snips (Fig. 4-5), commonly called aviation snips, use a system of links between the handle and the blades to create higher cutting forces, making them easier to use. Like conventional snips, they are available with straight or curved blades, handed left or right. The jaws of aviation snips have small serrations that help grip the material as it is being sheared (Fig. 4-6), making it easier to cut to a line. The disadvantage is that this leaves a matching pattern of serrations on the cut edges (Fig. 4-7), which may be undesirable on some work.

The Monodex Cutter

On thin sheets, the Monodex cutter (Fig. 4-8) uses a different action to reduce distortion. The top of the sheet is supported by the upper jaw of the tool, and squeezing the handles forces a pivoted blade through the sheet; this shears the sides of a thin strip from the sheet. The length of cut is short, and the cutter 'nibbles' its way along the cut. The cutting action removes a thin strip of metal, typically 2.7mm wide, and because the upper jaw of the cutter supports the metal, distortion is much reduced. There is a version of this tool specially for cutting corrugated sheet, in which the flat soleplate has been replaced by a curved sole designed to fit the curve of the corrugations. This is ideal for cutting roofing material.

Cutting Sheet Metal • 39

Fig. 4-8: EDMA Monodex nibbler shears.
EDMA OUTILLAGE

Fig. 4-9: Powered shears.

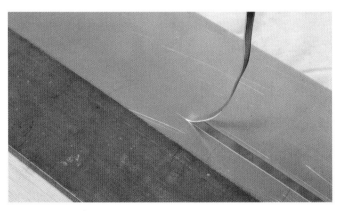

Fig. 4-10: Powered shears remove a strip from a sheet.

Powered Shears

Powered shears are available (Fig. 4-9), and the best of these removes a narrow strip (usually approximately 5mm or ³⁄₁₆in wide) from a sheet (see Fig. 4-10). As with the Monodex cutter, the upper soleplate of the shears has two cutting edges, and the moving blade (see Fig. 4-11) shears two edges simultaneously, removing a strip of material. The cut is advanced simply by pushing the cutter forwards. This is an ideal

Fig. 4-11: The blade of powered shears.

tool for cutting long lengths of material as the length of cut is unlimited. The blade moves up and down, typically at around 4,500 strokes per minute, and the cutting action is rapid and smooth.

This tool features an internal guide strip to make the cut strip curl to one side of the tool as it moves forwards. This is not always ideal, though, as the curl tends to twist downwards in a tight coil, marking the sheet as it twists. A better solution might be an arrangement to guide the strip further upwards before it coils, keeping it above the surface of the sheet. On a long cut, the cut may need to be paused at intervals so that the coil can be snipped off.

In use, the tool can be started with the front of the soleplate just resting on the sheet. Push forwards to start the cut, then gently tilt the body of the tool very slightly up or down until a 'sweet spot' becomes apparent and the tool glides smoothly and quickly forwards. The particular strength of this tool is its ability to follow a straight line with ease. Gentle curves are also relatively easy to follow; however tight curves are more difficult because the cutting blade is straight.

The edge of the soleplate can be run along a thin guide strip clamped to the work to provide accuracy with speed (see Fig. 4-12).

Note that although this tool cuts a strip from the sheet, it does not terminate the cut at the end of the strip, because the soleplate is not closed at the front. This means the tool cannot produce a finished end to a slot. Drilling a 5mm diameter hole at the intended end of the cut allows the sheared strip to fall off as the blade enters the hole (see Fig. 4-13).

Shears cannot begin a cut away from an edge, so cutting a shape within the boundaries of a sheet requires a pilot hole large enough to accommodate the moving blade: with the shears shown, this is approximately 45mm. While it is best to begin a cut at right angles to a straight-edge, the shears will

Fig. 4-12: Running shears along a side guide clamped to a sheet.

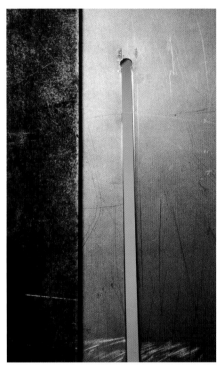

Fig. 4-13: Ending a cut at a punched hole.

begin a cut on a curve, if firmly guided and moved forwards slowly at first.

Hacksaws

Straight cuts can be made with a hacksaw, provided the blade is chosen carefully. As a general rule of thumb, there should be approximately three teeth in contact with the work at all times, so a fine blade will be required, with thirty-two teeth per inch (tpi) being useful for sheets of 0.8mm and thicker. Because the blade is normally used at an angle to the thickness of the sheet, 32tpi is useful from around 0.5mm thickness, with care.

Hacksaw blades should be tensioned in the frame, and the teeth should point forwards, away from the operator, so that cutting takes place on the forward stroke.

The snag with a conventional hacksaw is that it has a throat depth of perhaps 90–150mm, depending on the angle between the blade and the work, which restricts the distance it can cut into a sheet. When using a hacksaw to cut sheet, the edges on either side of the cut need to be supported against the downward pressure caused by the cutting action. Lay the sheet on a flat bench with the line of the cut close to the edge of the bench. Support the other edge on a roller stand or a folding workbench or some other similar arrangement. When cutting a thin sheet, there is a tendency for the sheet to be dragged upwards as the hacksaw blade is pulled back, and it may help to sandwich the metal sheet between two other sheets of hardboard, MDF or plywood, clamped close to the line of the cut.

The Sheet Saw

The sheet saw, panel hacksaw or Shetack saw is a cross between a conventional saw and a hacksaw, with a deep sheet behind the hacksaw blade, acting as support instead of the tubular frame and allowing cuts of unlimited length. This kind of saw is no longer in production, although they do occasionally come up for sale.

Pad Saws

Pad saws (Fig. 4-14) consist of a handle with a slot to hold a projecting hacksaw blade (or a shortened blade, which works best), and these can be used to make long straight cuts, if the saw is handled firmly and carefully. Mounting the blade with the teeth pointing towards the handle means that the blade is under tension when cutting, and this may help to prevent it from breaking. Cutting then takes place on the 'pull' stroke, which tends to lift the sheet upwards, so that clamps are essential. Keeping the blade upright while cutting helps, but wobbling to the side tends to snap the blade. Powered versions are useful in confined spaces (see Fig. 4-15).

Modern-day industrial electric saws can be fitted with a fine-toothed metal cutting blade and used instead of a hacksaw or panel saw, to good effect. They have no restrictions on throat depth or length of cut,

Fig. 4-14: Pad saw.

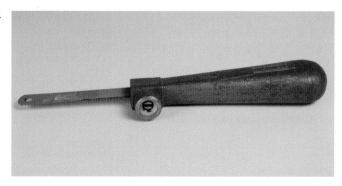

Cutting Sheet Metal • 41

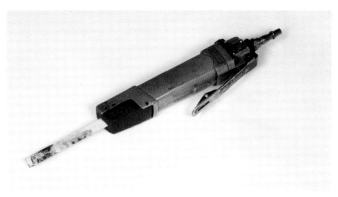

Fig. 4-15: Air-powered saw.

but do require firm guidance. Note that the relatively coarse blades of small consumer DIY saws make them unsuitable for cutting sheet metal.

Bench-mounted Shears and Cutters

Bench-mounted shears (Fig. 4-16) use a scissor action, like hand-held shears, but the long lever and the sturdy blade make cutting relatively easy. This kind of shear suffers from the disadvantage that cut material is likely to distort as it slides along the sides of the shears behind the blade. As with hand-held shears, the action of cutting will set up stresses on either side of the cut, and the resulting edges will tend to distort.

The Gabro bench-mounted cutters (Fig. 4-17) act in the same way as the Monodex, removing a thin channel of material by shearing on either side of the blade. These machines are supplied with their own tables and stands, so act as free-standing cutting stations. Blades vary in thickness along with the capacity of the machines, removing widths of 3.3mm or 4.5mm from sheet thicknesses up to 1.6mm or 3.2mm steel respectively. Cut lengths on the two current models are 70 or 108mm per stroke.

By removing a thin strip of material, this design of cutter avoids distorting the sheet, and allows long cuts with repeated strokes of the blade, as the sheet is free to pass on either side of the blade. Unlike powered shears, this tool does produce a finished end to a slot.

Sheet metal Guillotine

The sheet metal guillotine (Fig. 4-18) shears an entire edge using a long blade acting against the rear edge of a table, and can be operated by foot (in the smaller sizes), by an electric motor, or by pneumatic pump. Although this is a shearing action using what is effectively a large pair of scissor blades, cutting right across a sheet helps keep the cut edge straight and undistorted. The length of cut is limited to the length of the blade, but guillotines are available in large sizes, so that for the industrial user it is simply a matter of choosing the appropriate size of machine and capacity of cut to suit the purpose.

A sheet is placed against the side guide, which is at right angles to the blade, and the line of the cut is aligned with the cutting edge of the table. As the machine is operated, a clamp presses down on the sheet to hold it in position, and the blade shears the sheet from one side to the other. The machine can be fitted with a rear stop for repetition work, so that individual sheets do not need to be measured and marked but can simply be aligned with the side guide, pushed into the machine until they reach the rear stop, then cut.

Fig. 4-16: Manually operated bench-mounted shears.

Fig. 4-17: RMT-Gabro 2M2 notcher.
THE M J ALLEN GROUP OF COMPANIES

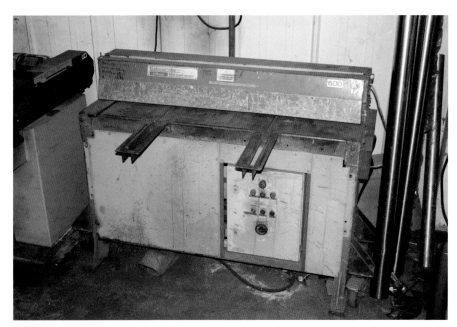

Fig. 4-18: Powered guillotine.

Fig. 4-19: Corner notching tool for a fly press.

Fig. 4-20: Self-contained corner notching punch and die set for use in a hydraulic press. The notch can be cut the full length of the aperture.

Miniature versions of this machine can be fitted to a fly press or a short hydraulic press. These have a much shorter cutting edge (for example, 100mm in a fly press) but the same cutting action. Operating the press forces a blade down flush with the end of a table or short support, and shears the material.

The Corner Notcher

One frequently required operation is cutting a square or rectangular notch from the corner of a sheet, in preparation for folding to form a box side or similar shape. The corner notcher is a short guillotine with two blades at right angles and it makes two simultaneous cuts, forming a corner cut-out in one operation. Although relatively expensive, this tool does exactly what is required in one easy operation, and in a production situation is a valuable asset. Fig. 4-19 shows a corner notcher for use in a fly press, but versions are available for use in a hydraulic press (Fig. 4-20) or as a stand-alone machine. Versions are also available to cut single internally or externally radiused (curved) corners (Fig. 4-21).

Fig. 4-21: External and internal radiused corners.

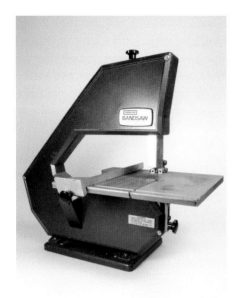

Fig. 4-22: Small bandsaw.

A Vertical Bandsaw

A vertical bandsaw can be used to cut sheet accurately and conveniently. This kind of saw uses a blade in the form of a continuous loop carrying teeth, rather like a longer version of a jigsaw blade, but much deeper back-to-front. The saw has a horizontal table and the blade passes through a slot, cutting as it moves downwards. The sheet is fed horizontally into the blade, with the cut edges passing on either side as the sheet is cut. This means any length of sheet can be cut without binding or being obstructed. A guide arm can be used to create a straight cut a defined distance from one edge of the sheet.

Although the blades are usually relatively deep, large radius curves can be cut, with care. A limiting factor is the depth of the throat on the saw (the distance from the side of the blade to the first point of contact with the inner side of the frame of the saw).

Bandsaws are commonly used for cutting wood, but the speed of travel of the blade on a metal-cutting bandsaw is significantly slower than for a wood-cutting saw.

Typical speeds for cutting steel are approximately 100 metres per minute (m/min) (300 feet per minute, or ft/min) for low-carbon steel sheet, 80m/min (250ft/min) for brass, and 130m/min (400ft/min) for aluminium. Slower speeds are required for harder materials, and a typical multispeed bandsaw might feature speeds from 20m/min to over 250m/min for cutting metal, and up to 400m/min for wood.

Tooth design for metal and wood is different, so a metal-cutting blade is required. Cutting metal with a wood-cutting blade will blunt that blade very quickly.

Points worthy of attention when choosing a metal-cutting vertical bandsaw are:

- The quality of the blade guides both above and below the table. Wobble in the blade tends to produce a wavy-edged cut. Good guides tend to be expensive, but are a worthwhile investment
- The size of the gap at the sides of the blade where it passes through the table. Sheet metal, and in particular thin sheet, requires support as close to the edge of the blade as possible. Inserts are often available to fit the hole in the table, and can be cut so that the slot clearance at the sides of the blade is small, which helps support the sheet at the point of cutting, producing a cleaner cut
- A cutting guide arm which clamps firmly to the table, to help prevent the edges of thin sheets becoming wedged under the guide

Be aware that bandsaws are a frequent source of serious injury. The appreciable length of exposed blade, and the fact that the operator is constantly feeding material towards the blade, means there is a high risk of contacting the blade. Bandsaws are often used in abattoirs and industrial butchers, and are particularly effective for cutting bone…

Take the time to adjust the top blade guides down as far as possible towards the top of the sheet, minimizing your exposure to the blade – and always retain a healthy respect for this machine.

Circular Saw Blades

Specially designed circular saw blades are available for cutting sheet metal, either in a saw table or a hand-held circular saw. These blades are normally tungsten carbide tipped (TCT) but with a specially designed tooth shape (the 'ATP' profile). Material such as 3mm aluminium cuts easily, but other metals can be cut too. Be aware that cutting thin sheet on a table saw is fraught with danger, as thin sheet is prone to slip under the side guide arm of the saw and become trapped. Any reaction from the saw is likely to throw the sheet straight back at you, so this method of cutting metal is not recommended. I strongly advise against it.

STRAIGHT CUTS AND TIGHTER CURVES

Nibblers

A powered nibbler (Fig. 4-23) uses a shearing action, but instead of removing a thin strip, it uses a punch (Fig. 4-24) to remove a small circle. By punching rapidly (typically 1,300 to 2,200 strokes per minute) a channel is nibbled from the sheet. Smaller punch sizes (typically 2mm) are used for thinner material (up to 1.6mm) and allow curves of radius approximately 50mm to be cut, while larger punch sizes (typically 4mm) are used to cut thicker sheet (up to 3.2mm) but are restricted to a minimum radius of approximately 128mm.

Nibblers leave a slightly serrated edge, jagged where one circular cut overlaps another, but the high speed of punching means the serrations are tiny. Take care, though – they are sharp. The faster the tool is pushed through the work, the larger the serrations, while moving the tool forwards

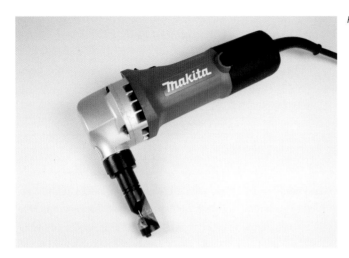

Fig. 4-23: Nibbler.

Fig. 4-24: Punch on a nibbler.

slowly creates much smaller serrations, which look increasingly like a smooth edge, although this is still sharp. Nibblers produce waste in the form of a large number of tiny crescent-shaped pieces of sheet, and these too are sharp, so they should be swept up and not handled.

In contrast to the powered shears, the nibbler is much more capable of following a curve. Its ability to maintain a straight edge depends on the skill and steady hand of the operator, and it is much easier to run the edge of the cutting column along a guide to produce a straight edge. Because the cutting action involves a punch passing through the material, the powered nibbler produces a finished end to any slot it cuts, although that end will be a semi-circular shape because the punch used is circular.

Fig 4-25 shows a nibbler held in a purpose-made stand to allow accurate and speedy cutting of circles. The base which supports the sheet being cut is set level with the upper face of the anvil in the nibbler (located in the lower part of the column). Although this particular adaptation is used for cutting short curves, the length of the cut is unlimited, and the addition of a simple straight guide bar at the side would allow straight cuts to be made with ease. A larger base would provide support for larger sheets.

Powered nibblers cannot start a cut away from the edge of a sheet, so cutting a closed shape like a circle within a sheet requires a pilot hole to be cut first, before the nibbler can be lowered into the hole to begin cutting. A pilot hole should be at least equal to the minimum radius required by the nibbler, and preferably larger.

Electric Jigsaw

An electric jigsaw can cut metal sheets varying in thickness from approximately 1mm to over 6mm (Fig. 4-26). Because jigsaw blades are narrower (teeth to back of the blade) than a hacksaw blade, the jigsaw can cut tighter radii, smaller circles, and more complex shapes. They can also make long uninterrupted cuts, either straight or curved, because the blade is suspended unobstructed below the flat foot of the jigsaw. The blades usually cut on the up stroke, and the foot helps prevent the sheet rising up as a result of the cutting pressure. It is important, though, to press firmly downwards while cutting, and not to press forwards too heavily, otherwise teeth will tend to catch and lift the jigsaw off the

Fig. 4-25: Nibbler on a stand, for cutting external circles.

Cutting Sheet Metal • 45

Fig. 4-26: Jigsaw.

Fig. 4-27: Card on the base of a jigsaw, to protect the metal being cut.

Fig. 4-28: Piercing saw.

sheet. When cutting brass, or any metal with a finish on the top face, glue card on to the foot, to avoid scraping the sheet (Fig. 4-27).

Although jigsaw blades have a tendency to wander, straight cuts can be carried out by sliding the edge of the jigsaw foot against a straight-edge guide clamped to the sheet.

The same guidelines on tooth pitch apply to a jigsaw as to a hacksaw. Ideally there should be three teeth in contact with the thickness of the sheet, so for most sheet this means a fine blade. Blades for cutting wood are not sufficiently hard for cutting metal, so a metal-cutting grade should be used. One exception to these rules is that good quality wood-cutting blades may be used for soft metals such as aluminium and brass, although the blade will have a short life. Aluminium also benefits from a slightly coarser blade because it tends to clog fine teeth.

There is a balance to be struck, though, between coarseness of the blade, ease of cutting, the tendency for a coarse blade to snag and force the jigsaw upwards, and the quality of the finished cut edges. Cutting brass requires sharp blades at all times.

The jeweller's piercing saw (Fig. 4-28), handled carefully, can make cuts of almost any shape in thin sheet. Blades range from 6/0 (finest) to No. 8 (coarsest), and they should be fitted to the frame with the teeth pointing downwards, towards the handle. Blades should be under considerable tension. In use, the saw is pulled gently downwards with just enough forward pressure to ensure the teeth are cutting.

PIERCING SAW BLADE GRADES

Piercing saw blades are coded by number, but each grade corresponds to a tooth pitch.

Table 7: Piercing saw blade grades

Grade	Teeth per inch	Tooth pitch (mm)
8	28	0.91
6	34	0.75
5	36	0.71
4	38	0.67
3	41	0.63
2	44	0.57
1	48	0.53
0	52	0.49
2/0	56	0.45
3/0	60	0.43
4/0	67	0.38
5/0	71	0.36
6/0	81	0.31
7/0	98	0.26
8/0	98	0.26

46 • Cutting Sheet Metal

Fig. 4-29: Bench peg.

To turn a corner, continue the up-and-down motion, but pause on the spot and gently rotate the saw or the work, letting the blade work its own way round. The best blades have rounded backs, to allow them to turn easily. Be prepared to change the blade as it loses its keen edge, or when it breaks. This is very much a tool that responds best to a practised touch. In use, the sheet needs to be supported close to the cut, and jewellers who frequently cut sheet use a bench peg – also called a bench pin – (Fig. 4-29), which is fastened to the edge of a bench and allows the material and the saw to be moved quickly to follow a curve while providing support for the cut.

The Fretsaw

The fretsaw (sometimes called a scroll saw) is a powered saw (Fig. 4-30) rather like a piercing saw, although with a more robust blade. The blade is powered up and down, and on some saws the stroke rate can be varied. Metals require a slower stroke rate than wood or some plastics. Fitted with a suitable blade for metal cutting, a fretsaw is capable of cutting intricate shapes. Using a hand-held piercing saw, both the saw and the sheet are turned during the cut. Using a fretsaw, the saw and blade remain stationary, and the sheet is turned to make the blade follow the line of the cut.

Fig. 4-30: Fretsaw.

Fretsaw blades for metal cutting vary from grade 0000 (70tpi) to grade 9 (25tpi). Blades often tend to cut towards one side, especially when the pace of cutting is forced, so, as with the hand saws, cutting should take place at a pace determined by the blade itself.

Blades should be fitted with the teeth pointing downwards, cutting speeds (that is, stroke rate) should be slow (and much slower than for wood), and where the saw has an adjustable stroke length, shorten the stroke, especially for thin materials, to help lower the rate at which blades break. Sheets can be taped to a backing board of plywood or MDF, or they can be sandwiched between two sheets of ply or MDF.

On the fretsaw, despite the direction of cutting being downwards, the blade tends to try to lift the sheet off the worktable on the back stroke. Resisting this requires downward pressure with the fingers, near the saw blade. Strong downward pressure will tend to make it difficult to maintain forward movement and fine control of the direction of cut, so there is a compromise here, best worked out with practice. Take care when your fingers are near the blade, and keep the plasters within reach.

DEBURRING EDGES

The edges of sheets tend to be sharp, and cut edges may be sharp, with needle-like slivers in places, so it is important to deburr edges to prevent injury. This also allows the edges to mate properly with adjacent parts, or to sit flat on other flat sheet parts. Keeping the cutting edges on your tools sharp is the first step.

Basic deburring tools include files, carbide wheel tools, belt sanders, and purpose-designed sheet deburring machines.

Files

Files can be used to deburr short lengths of sheet. File across the edge, at an angle of approximately 45 degrees (Fig. 4-31). Then make a few strokes along the sharp edge between the flat side of the sheet and the narrow edge, tilting the file at approximately 45 degrees to the flat side as well as 45 degrees to the vertical (Fig. 4-32). The thinner the sheet, the finer the file that should be used. Files come in many shapes and sizes, but an engineer's hand file, with a long, flat, rectangular blade, is best for filing

Cutting Sheet Metal • 47

Fig. 4-31: File an edge horizontally, at 45 degrees.

Fig. 4-32: File an edge at an angle of 45 degrees to the horizontal and 45 degrees to the vertical.

Fig. 4-33: Half-round and circular-section files.

Fig. 4-34: Deburring tool fitted with carbide wheels for deburring sheet edges.

bastard, second cut, smooth and dead smooth. When working with brass sheet you may wish to use a finer 'Swiss' pattern file. This is a hand file, but with a range of finer cuts from 00 (coarsest) through 0, 1, 2, 4, 6, and 8 (smoother) then 0/1, 0/2, 0/4, 0/6 and 0/8 (very finest). A 0 cut is relatively coarse, while an 8/0 will leave a very fine finish indeed. The coarser files are more suitable for quick deburring, but following this with increasingly finer files leaves a fine finish.

Half-round files, and files with a circular cross-section (Fig. 4-33), are suitable for deburring curved cuts and holes of various sizes.

Support any edge being filed to prevent distortion and noise, either by resting the sheet face down on a flat bench top, or by clamping the sheet between two bars placed close to the edge. Small pieces of sheet can be gripped between smooth jaws in a vice, gripping close to the cut edge to provide support.

Carbide Wheel Tools

Purpose-made deburring tools for sheet consist of two hardened wheels attached to a handle with a hand guard to prevent injury (Fig. 4-34). The wheels lie at right angles to the sheet, which is caught between the edges of the wheels (Fig. 4-35), and the tool is simply drawn along the cut edge of a sheet.

The edges of holes or curved cuts can be deburred, one side at a time, using a

a flat edge. For curved edges, a half-round file (flat on one side, semi-circular on the other) works well.

The coarseness of a file is called the 'cut', and engineer's files are available in cuts varying from coarse, through medium,

Fig. 4-35: Pull the tool along the edge of the sheet.

Fig. 4-36: Deburring tool with swivelling blade.

swivelling deburring tool with a hardened blade (*see* Fig. 4-36). Drawing the blade along the sharp edge makes it cut a thin sliver of metal, effectively bevelling the edge. This should be repeated on the adjacent corner so that both sides of the cut are deburred.

Belt Sander

A hand-held abrasive belt tool, such as a power file, can be used to deburr edges and corners quickly. Fit the tool with an appropriate grade of belt suitable for metal, such as one that uses an aluminium oxide abrasive. Using the same cuts as a file, the power file carries out the same job more quickly. Take care, though, to keep the belts moving, because dwelling for too long in one area may cause the belt to remove too much metal. This is a deburring operation, and the intention is not to grind the sheet to shape. Hold the file so that the belt is moving off the edge of the sheet and not towards the edge. That will minimize any tendency to snatch, catch or dig in.

Purpose-made Deburring Machines

Purpose-made deburring machines can be used in a production environment, and these powered machines use grinding wheels or flap discs to deburr one or both corners of the edge of a sheet both quickly and easily.

INDUSTRIAL PROCESSES

Sheet material can be cut by several industrial processes in which the machinery is often computer controlled or, in earlier forms, where the machine uses an optical sensor to follow a path on a drawing. These processes involve laser beams, plasma gas jets, or streams of high-pressure water mixed with abrasives.

Although mainly industrial machines, plasma cutters are readily available at reasonable cost, and laser cutters are almost within reach of the home or light industrial user, the limitations currently being on the type and thickness of material that can be cut. Wood and paper are easily cut, but metal requires much more expensive lasers of higher power. Because of the very high pressures used, water jet cutters are currently industrial-only processes, and are likely to remain expensive to buy and maintain. This does not preclude their use, though, and it is a relatively straightforward process to prepare a design to be cut by any of these processes by a specialist company.

One effect of this kind of cutting is that the design of sheet parts can often be altered to speed subsequent assembly. Because these cutting processes are capable of high accuracy, the accuracy of positioning of subsequent processes such as bending or punching may be improved. Some assembly methods, such as the use of accurate castellated joins, which may be difficult to achieve with other methods, become much easier to carry out simply because of the accuracy of fit of the individual parts.

Locations of holes can be marked by spotting pilot holes through the material, but one limitation is on the diameter of small holes in relation to the thickness of the sheet, and this varies depending on the cutting process to be finished later using conventional processes. Another option is to completely finish holes as part of the cutting process.

Laser Cutting

Laser cutting is quick and accurate and involves providing an accurate technical drawing and a specification for the material, to a company specializing in laser cutting.

Fig. 4-37: Laser-cut edge.

There will be an initial set-up charge, and an additional cost, which depends on the length of cut. Parts can be completely cut out of the sheet, or can be left attached to the sheet by small tabs which can be cut by other means later. Cut edges are normally accurate to within 0.05mm (0.002in), with slight tapering which is not noticeable on thin sheet (Fig. 4-37), and the edge may require a light clean-up and deburring on the bottom side. The cut edge will have been subjected to thermal stress, and steel edges may have been hardened by the process.

Water Jet Cutting

Water jet cutting involves directing a high pressure jet of water and abrasive particles at the sheet. Unlike laser cutting, water jet cutting is a cold process involving no thermal stress on the edges of the cut, and no burning of the edges. Tolerance is normally held within 0.2mm (0.008in). Water jet cutting becomes difficult with thinner materials, and it is perhaps best suited to materials of 3mm or greater thickness.

Plasma Cutting

In plasma cutting a jet of gas is directed at the surface of the sheet, and an electrical arc is passed from a cutting head to the sheet, through the gas. The gas is turned into a plasma (an ionized gas) and the heat melts the sheet, while the gas blows the melted particles through the sheet. Tolerance for plasma cutting is approximately 0.5mm (0.02in), so this is not as accurate a process as laser cutting. On modern 'high-definition' plasma cutters, it is claimed that the quality of finish on the cut edge approaches the quality of laser-cut edges, and needs little finishing. Plasma cutting is an effective way of cutting thicker sheet.

5 Making Holes

Holes vary in size and shape, as do the methods used to make holes in sheet metal. There is often more than one way to make any particular hole, and the method chosen usually comes down to the available equipment, the accuracy and finish required, and safety concerns.

CLAMPING

Because many of the methods for making holes involve rotating tools that are likely to snatch and grab sheet material, secure clamping arrangements are an absolute must when creating holes in sheet (*see* Fig. 5-1). Two clamps provide more security than one, preventing the sheet using one clamp as a pivot point as it breaks free from the drill. There are no exceptions to the use of proper clamps, as there are few things more dangerous than a loose sheet that is spinning freely around a drill bit or hole saw, slicing its way through everything it touches. A good set of clamps, used consistently, is an essential investment.

DRILLING

Small holes of 3mm and under can usually be drilled with a twist drill. For most metals, a jobber drill with conventional geometry works well, but there are some advantages in matching the geometry (or shape) of the tip of the drill to the material being drilled,

OPPOSITE PAGE:
Louvres on the side of the bonnet of a vintage car.

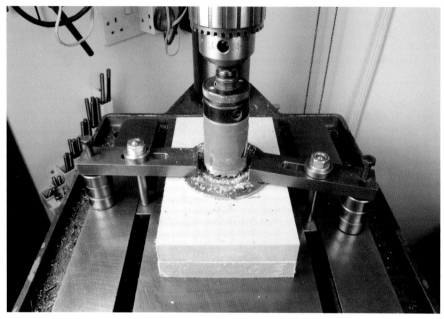

Fig. 5-1: Use at least two strong clamps to secure work in the drilling machine.

DRILL GEOMETRY

- Standard jobber twist drills have a tip angle of 118 degrees
- The angle between the cutting edge and the inside of the helical flute is called the rake angle
- The lengths of the two cutting edges should be equal
- Drills are available with slower (more vertical) flutes or quicker (more twisted) flutes

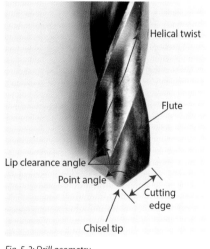

Fig. 5-2: Drill geometry.

whether this is mild steel, stainless steel, brass or copper.

Mild steel: Use standard drill tip angles (normally 118 degrees included angle, as found on a jobber drill).

Stainless steel: This material has a tendency to work harden because of the heat generated during the drilling process, making it progressively more difficult to feed the drill into the work, especially when drilling deeper holes, so use a standard drill geometry, firm pressure, and a medium to fast drilling feed rate. Consider using a solid carbide or carbide-tipped drill – for example a 'masonry' drill – if available, and a coolant such as neat cutting oil.

Brass: Conventional jobber drills tend to snatch and draw themselves into the work, or draw the material up the drill bit, especially as the drill exits the hole. Purpose-designed drills exist although they are not always readily available, but good results are obtained by modifying a conventional jobber drill by backing off the angle at the cutting edge of the drill tip.

Copper: Twist drills with conventional geometry tend to catch badly in copper, and there is a tendency towards seizure of the drill in the material. Copper chips tend to stick to the tip of the drill at the cutting edge, making it difficult to drill a clean hole with a decent finish. As with brass, alter the cutting edge, giving it a zero or negative cutting angle. Use a cutting fluid such as Castrol Ilocut 486, Relton A9, or WD 40.

Use a TiN-coated cobalt or carbide drill (but do not alter the geometry of the cutting edge, as this will remove the TiN coating).

Note that parabolic drills, which have a curved cutting edge and a different shape of flute, are not suitable for drilling sheet as they have a greater tendency to snatch the material.

Keep your drills sharp!

Twist Drills

On larger holes, conventional twist drills snatch and grab, and they tend to produce a non-circular lobed hole (*see* Fig. 5-4) because deflection of the thin sheet under the pressure of the cutting edges makes the drill wobble. The sheet may also distort as it bends and stretches under the pressure of the cutting edges. All of this makes a conventional twist drill virtually useless for drilling holes larger than 3mm diameter. If you must use a larger drill, use the appropriate top speed for that drill, and don't force the feed.

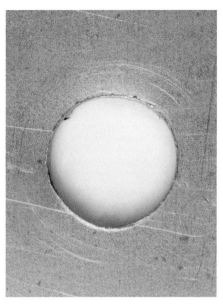

Fig. 5-4: Lobed hole.

BACKING OFF THE RAKE ANGLE

The rake angle is the angle of the flute at the cutting edge. For drilling brass, reduce this to zero by using a small slipstone (grinding stone) to make the flute vertical for a tiny distance at the cutting edge. Hold the drill vertical, hold the slipstone vertical, and rub the slip against the cutting edge where it meets the flute until there is a small flat. Repeat for the other cutting edge and flute.

When the drill is used, and the cutting edge penetrates the work, it pushes that short vertical section into the work.

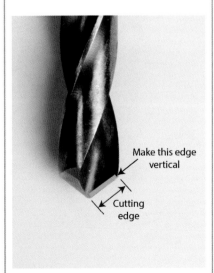

Fig. 5-3: Backing off a drill rake angle at the cutting edge.

DRILLING SPEEDS

The table below shows recommended drilling speeds using good-quality high-speed steel twist drills (jobber drills) in a substantial drilling machine. Speeds may be a little lower, but no faster.

Table 8: Drilling speeds

HSS twist drill diameter	Drilling speed (rpm)	
	Mild steel	Soft aluminium
1	10,000	30,000
2	5,000	15,000
3	3,000	9,000
4	2,500	7,500
5	2,000	6,000
6	1,600	4,800
8	1,250	3,800
10	1,000	3,000
12	800	2,400

Making Holes • 53

Fig. 5-5: Butterfly drill point.

Fig. 5-7: Hole saws.

Butterfly Drill

A butterfly drill has a small conical centre point, and the cutting edges are two lips at the outer circumference of the drill (Fig. 5-5). The centre point locates in a centre pop, steadying the drill, and cutting takes place at the outer tips. The drill effectively cuts a circular disc from the parent material, leaving a clean hole with a circular edge. The conical centre point does not cut, but presses into the sheet, distorting it. However, that centre pop is in the waste portion which is removed to create the hole, so distortion in that part is not important (see Fig. 5-6).

The Hole Saw

The principle of cutting on the circumference is the basis of the hole saw (Fig. 5-7), which acts like a hacksaw blade bent into a circle. A central pilot twist drill creates a hole, and is then used as a steady to guide the saw blade into the work. Cutting takes place at the circumference, but the action is more like a saw, with the geometry of a saw tooth rather than a drill bit.

The accuracy produced by a hole saw is not great, as the saw blade tends to wobble slightly. For good support from the central pilot drill, that drill bit really needs to be located in hard or firm material beneath the sheet and tightly clamped to the sheet.

Hole saws need to have a tooth pitch appropriate to the material being cut, with a relatively fine tooth pitch for thin sheet, and be run at a speed appropriate to the diameter of the saw.

Step Drill

Using the principle of a central support, the step drill consists of a series of cutting edges, each with a very short section of full

Fig. 5-6: Waste removed by a butterfly drill.

HOLE SAW CUTTING SPEEDS

The table below shows recommended maximum cutting speeds using good-quality bimetal hole saws incorporating a pilot drill, in a substantial drilling machine. Speeds may be a little lower, but no faster. Rigidity is essential, as is secure clamping of the work. A lubricant is recommended: cutting oil for steel, white spirit or a proprietary lubricant for aluminium.

Table 9: Hole saw cutting speeds

Bimetal hole saw diameter	Drilling speed (rpm)	
	Steel	Aluminium
14	580	900
16	550	825
19	460	690
20	440	650
22	390	580
24	370	555
25	350	525
30	270	400
40	220	330
50	170	260

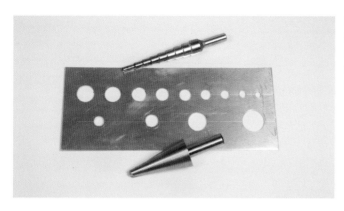

Fig. 5-8: Step drill (above plate) with drilled holes (upper row), and conical drill (below plate) with drilled holes (lower row).

Conical Drill

A conical drill, often referred to as a tube and sheet drill, or by a proprietary name such as Conecut, has a conical body whose diameter varies continuously; unlike the step drill, it does not have discrete steps or diameters (*see* Fig. 5-8). The point of the drill acts as a pilot and may have a fixed diameter for a short distance behind the point. Thereafter, the drill produces a hole whose size increases as the drill is fed into the work. The sides of the hole will be tapered, matching the taper of the conical drill, although for many jobs this is unimportant. Where a hole must have parallel sides, use a step drill or some other method of producing the hole.

PUNCHING

Clean holes of accurate size in sheet material are perhaps most often and most conveniently produced by using a punch and die.

diameter flute, stacked so that each size provides support as a pilot for the next size (*see* Fig. 5-8).

Drilling begins with the centre point, and progresses by successively enlarging the hole until it reaches the desired diameter. Clearance must be provided to allow the steps to protrude beneath the sheet, but this kind of drill provides a very practical way of drilling large, clean holes in sheet material. A drilling speed should be set that is appropriate to the diameter of hole being drilled at each stage, although in practice it is most convenient simply to set the drilling speed for the largest diameter of hole that will be drilled. When drilling a specific diameter, put tape around the next larger step, as a visual reminder of when to stop.

A punch is a piece of material whose cross-section is the same as the hole it will produce. A die is a piece of material with a hole shaped to receive the punch. In use, the punch is held directly above the die; the sheet is placed between the punch and the die; then the punch is pushed through the sheet, passing into the die (*see* Fig. 5-9). The punch shears the sheet material, and the edges around the punched hole are supported by the die, so shearing takes place within the sheet between the edge of the punch and the edge of the hole in the die. There is normally a small clearance between the outside edge of the punch and the inside edge of the die, and a taper away from the cutting edge within the die, to help the cutting action, but punching is quick and accurate and normally produces a clean, finished edge in the material.

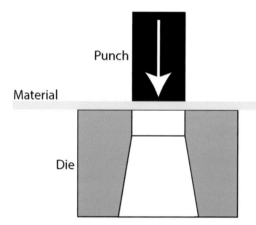

Fig. 5-9: Punch and die.

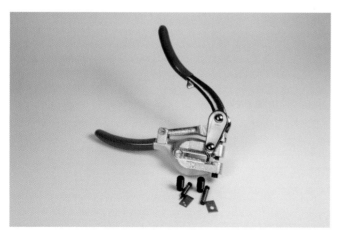

Fig. 5-10: Hand-held punch.

Holes up to 7mm diameter can be punched in sheet steel up to 2mm thick by using a hand-held punch (Fig. 5-10) with a compound action handle, a small toggle

Making Holes • 55

Fig. 5-11: Small toggle press used to punch holes.

Fig. 5-12: Small hammer-operated punch, fitted with a width guide.

Fig. 5-13: Small arbor press.

Fig. 5-14: Bolster, die and stripper plate mounted on a small arbor press.

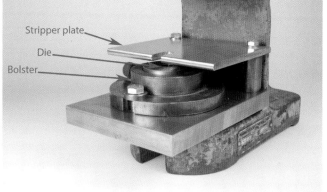

Fig. 5-15: Punches often have tilted cutting edges to pierce the work and lower the cutting force required.

punch is secured in the ram, and the die is held on the foot of the press. Various accessories can be attached to the press, to hold and guide the work and allow precise positioning of holes. Small punches may have flat ends at right angles to the axis of the punch, but larger punches benefit from a tilted face so that the work is pierced at one edge of the hole and progressively sheared around the edge as the punch enters the sheet (Fig. 5-15).

Self-contained frame-mounted punches (Fig. 5-16) are designed to be operated by a small hydraulic press or in the jaws of a press brake, and large frame-mounted punches, such as the Redman series (Fig. 5-17), can punch a range of shapes and large diameter holes, but may require a more powerful 50-tonne press.

A fly press (Fig. 5-18) performs much the same basic action as a lever-operated press, but is capable of greater forces and a faster action. The fly turns the screw, which pushes

press (Fig. 5-11), or a hammer-operated punch (Fig. 5-12), but the force required to shear metal increases with the thickness of the sheet and the diameter of the hole, so larger holes, thicker sheet, or the stress of repetition demand a different approach.

Mounting a punch and die set in a manual or hydraulic press allows greater force to be applied, as well as guiding the punch accurately. Small lever-operated arbor presses (Fig. 5-13) can exert pressures of a half to two tonnes, depending on the model and size of press, and these can be adapted to carry punches and dies, as shown in Fig. 5-14. The

Fig. 5-16: C-frame punch.

Fig. 5-17: Redman punch, designed to be used in a powerful press to punch large holes.

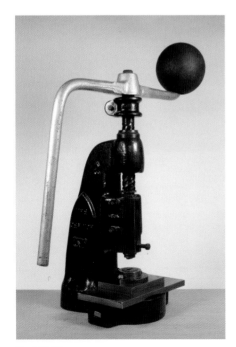

Fig. 5-18: Fly press.

so that the sheet can be slid into place easily, and just as easily removed, but the stripper is sufficiently close to the sheet to stop it so that the punch may withdraw.

In addition to plain holes, a range of shapes can be punched to create internal holes. Fig. 5-19 shows a tool for punching a rectangular slot. The tool has an extra protruding guide which remains in the die to guide the cutting edges to ensure alignment as the punch enters the die, and the die has two flats to orientate the die within a large bolster. Fig. 5-20 shows a circular punch to create a large hole, and Fig. 5-21 shows a small range of punches and dies (not all matching), including a corner-notching punch and die (on the right). Fig. 5-22 shows that the circular portion of the punch locates in the curved section of the die.

Blanking operations can also be carried out. Blanking is similar to punching, but the punches and dies are designed to punch a workpiece from a sheet. Whereas the ram downwards. A multi-start screw is normally used, so that the ram descends a greater distance as the screw turns. Securing a punch in the ram and a die resting on the foot of the fly press allows holes to be punched in sheet material very effectively. The die is often held in a bolster (see Fig. 5-14), which allows a range of dies with the same outside diameter to be gripped firmly, and the ram has a standard size of hole so that it can hold a range of punches with bodies that have the same outer diameter.

Additions include stripper plates, material guides, locating pins and clamps. After a hole has been punched there is a tendency for the sheet to stick to the punch and rise as the punch is retracted. To avoid this, a stripper plate is often used to hold the sheet down (see Fig. 5-14). There is a generous clearance between stripper plate and sheet,

Fig. 5-19: Punch and die for creating a rectangular slot.

Fig. 5-20: Punch for producing a large-diameter hole.

Fig. 5-21: Assorted punches and dies, including a corner notching set (at the right-hand end of the top and bottom rows).

Fig. 5-22: The rear circular portion of this corner-notching punch locates in the die and acts as a guide.

the material punched from inside a hole is waste, in a blanking operation the material punched from the hole is the workpiece. The consequence of this is that when punching, it is the hole diameter which is important, so the punch should be dead to size, and the hole in the die should be up to 10 per cent larger, to provide clearance. When blanking, the finished size of the blank is important, so the die is made dead to size and the punch is made up to 10 per cent smaller.

Aside from basic holes, successive punching or blanking operations can be used to produce additional holes in a workpiece. For this, some means of locating the blank accurately for successive operations will be required, and the exact design of this will vary to suit the shape of the blank at each stage.

Punching a row of rivet holes, for example, can be carried out using a guide rail or backstop (Fig. 5-23) to ensure all the holes are punched at the same distance from the rear edge. That same technique can be used with a hand punch, within the limits of its throat depth; Fig. 5-10 shows that the hand punch is fitted with an adjustable backstop. Similarly, the punch illustrated in Fig. 5-12 is fitted with a side guide to position a hole centrally across a strip, or a specific distance from an edge.

Using a similar idea, but locating from a hole instead of an edge, a washer can be produced by punching the centre hole, then using a pin projecting from a larger-diameter punch to locate the centre hole immediately before blanking the washer circumference (Fig. 5-24).

Using a louvre punch and die set, the punch pierces the sheet, then pushes the material into a shaped die, to produce a louvre (Fig. 5-25).

Larger holes can be produced by multiple punching operations, so that a large rectangular hole, for example, might be produced by punching several smaller holes, as shown in Fig. 5-26. This requires appropriate stops to position the workpiece, but can result in a well-finished hole. This is the principle behind some CNC punching machines, which can nibble a large hole using smaller punches.

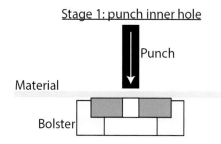

Fig. 5-23: A backstop may be used to locate holes from an edge.

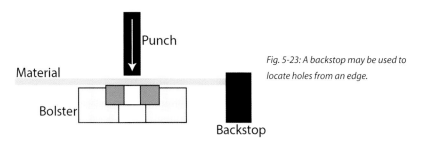

Fig. 5-24: Successive punching and blanking operations to produce a washer.

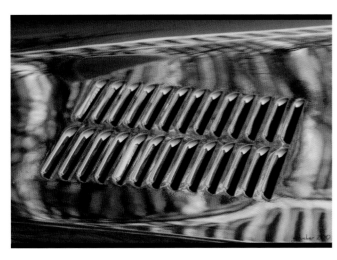

Fig. 5-25: Louvres on a car bonnet at the Mt Clemens Cruise Night. GARY TUCKER

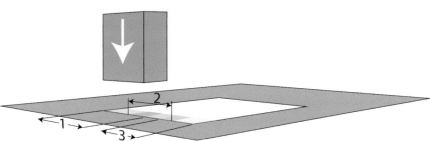

Fig. 5-26: A large hole can be produced by punching a pattern of smaller holes.

Fig. 5-27: Q-Max punches are operated by turning a screw.

Fig. 5-28: Portable hydraulic hole punch.

The Q-Max sheet metal punch uses the punch and die arrangement of the fly press, but applies the shearing force by means of a screw that can be tightened by hand. There are two piercing edges on the punch (Fig. 5-27), so it shears from two opposite sides simultaneously, helping prevent any tendency for the punch to tilt as it cuts through the sheet. Drill a pilot hole to allow the screw to pass through the sheet, assemble the punch and die on opposite sides of the sheet, then turn the screw until the punch has passed right through the sheet. This will produce a neatly cut hole in the sheet. The punch side will tend to be dished at the edges slightly, while the die side may have a small protruding burr which can be filed or ground off where necessary. Round, square and rectangular punch and die sets are available in a wide range of sizes.

Portable hydraulic hole punches of this sort are available to deal with the often considerable force required to operate a large hole cutter (Fig. 5-28).

TREPANNING

Larger holes can be produced using a trepanning tool (Fig. 5-29). This tool is best used in the drill press. The cutting tool can be offset a specific distance equal to the radius of the circle; a pilot hole is drilled,

Fig. 5-29: Trepanning tool. Versions are also available with plain shanks to suit a drill chuck.

then the cutter is lowered slowly into the work, rather like using a hole saw, and the circle is cut through at its circumference. The trepanning tool must be run at an appropriate speed; for a large-diameter circle, this is very slow. It is also prone to snatching, so the work must be securely clamped in place, supported on a firm backing plate. With care, this is a useful method of producing large-diameter holes, because the diameter of the hole can be set quite easily.

MILLING HOLES

Holes can be produced using a milling machine, and an end mill or slot drill (Fig. 5-30). Mounting the sheet on a rotary table with the cutter offset from the centre of rotation allows large holes to be cut easily, up to the maximum size of workpiece that can be swung around on the table (Fig. 5-31). Feed the cutter slowly into the work as the table revolves, increasing the cut slowly, over several revolutions if necessary. Finish the cut with the cutter well through the work, and carry out a full revolution at that depth of cut, to ensure a good finish on the edges of the hole.

On a CNC mill, a rotary table is not required. Place the sheet on to a firm backing on the mill table, and clamp securely. Set the controlled point to the centre of the circle, and use appropriate G2 or G3 commands to make the X and Y slides move in a coordinated path to cut a circle. Finish in the same way as for a manual machine, with a final cut to tidy up the inside edges. It is also possible to cut the circle a small amount (0.2mm or so) smaller in diameter than is required, then take a final full-depth cut at the final size to clean up the edges.

Fig. 5-30: Machining a circular component with a central hole from a 6mm aluminium sheet using a milling machine.

Fig. 5-31: Sheet component mounted on a rotary table in the milling machine.

6 Bending Sheet Metal

Sheet metal has mechanical properties that determine how it behaves when it is subject to force. Taking a long, narrow piece of thin, mild steel sheet in your hands and applying pressure to bring one hand a little closer to the other will cause the sheet to deform (change its shape) and curve. Releasing one hand allows the sheet to return to its original flat shape because the deformation caused by bringing your hands together was not permanent. But taking a short section of sheet and bending it forcefully and sharply between your thumbs, then releasing it, is likely to result in a permanent bend in the sheet. Even though the sheet might spring back a little when you release the pressure, there is some permanent deformation and the sheet permanently changes its shape. So why does this happen?

Sheet metal has a degree of elasticity, or a tendency to spring back to its original shape, but if it is subjected to a force that takes it beyond its elastic limit, it deforms permanently. We can use this behaviour to change the shape of the sheet and make it into, say, a box or some other useful shape. Applying a force to the strip causes stress (load per unit area), and that stress causes strain (deformation). Stress and strain do not necessarily cause *permanent* deformation; they are simply the applied force and the change in shape at any moment, compared to the unstressed condition when no force is applied.

OPPOSITE PAGE:
GABRO BF-1000 box and pan folder.
THE M J ALLEN GROUP OF COMPANIES

Metals have a modulus of elasticity, which is a measure of the tendency of the material to deform under stress but return to its original shape when the stress is removed. Permanent deformation does not occur until the stress reaches the elastic limit for the material. Applying greater stress then causes permanent deformation. So bending mild steel sheet into a gentle curve is unlikely to create enough stress to exceed the elastic limit for that material, and the sheet springs back when you release the force.

Bending the sheet more forcefully into a tighter curve between your thumbs creates much more stress and is likely to exceed the elastic limit for the material, resulting in some permanent deformation. The material does not collapse suddenly as soon as its elastic limit is reached. Instead, the amount of deformation is related to the amount of stress beyond the elastic limit, although this is not necessarily a linear relationship where twice the stress causes twice the deformation, three times the stress causes three times the deformation, and so on. It is enough to know that in order to cause permanent change of shape, the metal needs to be subjected to enough stress to exceed its elastic limit.

All of this is important because our ability to make a permanent bend in a sheet depends on the stress we cause by applying force, the modulus of elasticity, and the elastic limit of that material. Sheet metal suppliers can provide information on those properties of a sheet of any particular material so that engineers can calculate the forces required to cause permanent deformation in the sheet. That information will include the modulus of elasticity and mechanical properties such as the proof stress, tensile strength, shear strength, hardness and elongation, all of which are useful for calculating the behaviour of the sheet as various operations take place.

As bending takes place, a flat part is subject to considerable forces and bends which will stretch the material on the outside and compress it on the inside of the bend. That will have an effect on the length of material in the bend, and it adds a little complication when working out the 'flat pattern' or 'development' of the shape required to be cut from the flat sheet so that the final bent part will be a specific shape and size.

The ideal way to deal with developing a flat pattern is to use a CAD program capable of dealing with sheet metal, or the slightly more laborious manual methods and calculations.

This chapter deals with machines and methods of making bends, while Chapter 2 outlines the procedure for producing flat patterns and developments from a CAD program. Be aware, though, that you do need to start with an accurate flat pattern to end up with a final shape which is accurately to size.

BEND LINES AND BEND SIZES

Fig 6-1 shows three bend lines: bend line 1 lies at the start of the bend; bend line 3 lies at the end of the bend, while bend line 2 lies mid-bend. In a bend of constant

62 • Bending Sheet Metal

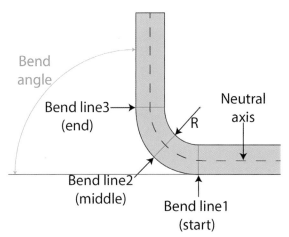

Fig. 6-1: Three common bend lines, and the neutral axis.

radius, it lies midway between bend lines 1 and 3.

The inside radius and outside radius are as shown. In this book, unless otherwise stated, the bend radius of a sheet metal bend refers to the inside radius.

The bend angle is the angle through which the material is bent.

Bend lines 1, 2 and 3, and the bend radius, all play their part in positioning flat work in a folder or sheet metal bender.

FOLDERS

In most cases, sheet needs to be bent into a particular shape, so bending between fingers or thumbs is not a reliable industrial method, and purpose-designed bending machines are normally used. The simplest bending machines apply force close to the bend, and, like bending between thumbs, apply force over a short distance and a narrow area of the sheet, to create enough stress to exceed the elastic limit of the material.

Fig. 6-2 shows a simple bending machine, usually referred to as a 'folder', but sometimes termed a 'bending brake': the difference is simply one of regional language. The material is clamped between two strong beams, and a third beam, mounted on a hinge, pushes the unclamped part of the sheet upwards (Fig. 6-3). The bending force is applied close to the clamp, to create maximum stress and to give a tight curve. The hinged bending beam can be rotated upwards to a suitable position to create the required angle of bend. When clamped, bend line 1 (Fig. 6-1) should be positioned at the front edge of the clamp beam.

More advanced folders have their top clamp beam tilted at an angle to accommodate a workpiece with a previously folded rear edge, and to allow adjustment of the clamp beam backwards, away from the edge of the lower beam (the 'bed') to allow for different bend radii; there is also a vertical adjustment to allow for different thicknesses of work. Fig. 6-4 shows a folder of this type, and Fig. 6-5 shows the adjustments available.

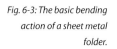

Fig. 6-3: The basic bending action of a sheet metal folder.

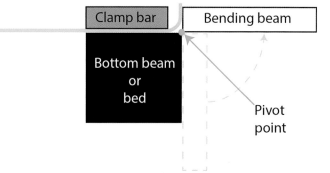

Fig. 6-2: A simple folder: the GABRO SF606.
M J ALLEN GROUP OF COMPANIES

Fig. 6-4: Folder featuring a solid clamp beam.

Bending Sheet Metal • 63

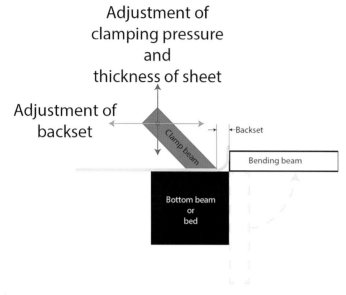

Fig. 6-5: Folders require adjustable backset and clamping pressure.

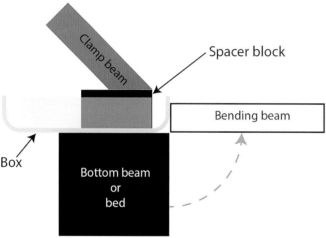

Fig. 6-6: Using a block under the clamp bar to avoid fouling the vertical sides of a box as the other sides are folded up.

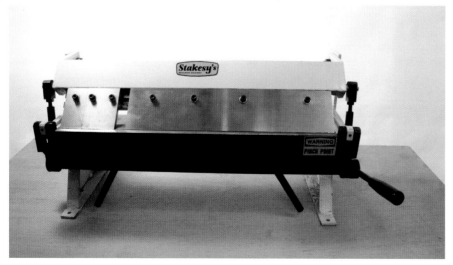

Fig. 6-7: A box and pan folder, with a segmented clamp beam comprising individual fingers of different widths.

Although the upper clamp beam is normally solid, there is often a gap at each end of the beam so that previously folded material can swing past the clamp beam without being obstructed. This is useful for some jobs.

Where a box is to be folded, and the sides would otherwise be obstructed by the clamp beam, some folders allow sufficient vertical adjustment to place a metal block under the clamp beam, allowing clearance for adjacent sides as the box is folded upwards towards the beam (Fig. 6-6). However, this requires care in aligning the block as the beam is clamped, before the bend is made. The clamping force should be sufficiently high to ensure that the block does not slip out as the bend is made and the block is forced backwards under pressure from the bending beam.

Box and pan folders are perhaps the most versatile type, as these have a segmented upper clamp beam that allows a box or similar object to be folded without the previously folded sides being obstructed by the rest of the clamp beam as they swing upwards and inwards (Fig. 6-7). The segments or 'fingers' are usually of different widths, and can be used singly or arranged in a row to create a range of bending lengths for different jobs. Removing fingers not required, or sliding them to the side to create gaps at the sides of the row, allows the length of the row to be matched to the length of the bend, while the gaps at the ends accommodate existing folded sides as they swing.

The finish on the surfaces of the clamp beam, the bed and the bending beam, as well as the more precise adjustments available, mean this type of folder is usually capable of more precise work.

CLAMP BEAM BACKSET

When making a fold, the thickness of the sheet (T) and the radius of the bend (R) both need to be taken into account

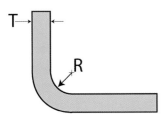

Fig. 6-8: When making a fold, the thickness of the sheet (T) and the radius of the bend (R) both need to be taken into account.

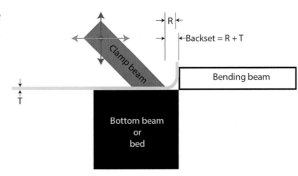

Fig. 6-10: The effect of R and T on the clamp beam backset.

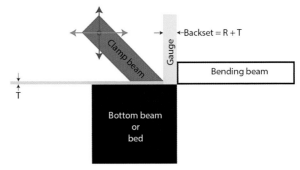

Fig. 6-11: Using a gauge to set the backset of the top clamp beam.

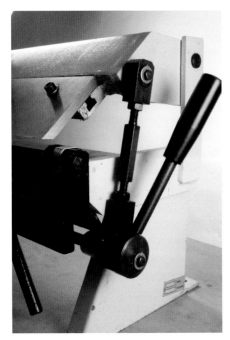

Fig. 6-9: Cam clamping mechanism to control clamping pressure.

Fig. 6-12: Using a piece of sheet as a gauge to set the distance between the bending beam and the hinge plate.

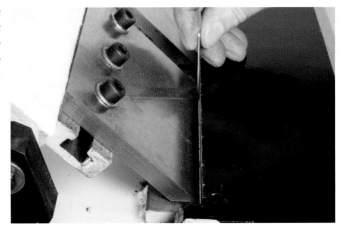

(Fig. 6-8). The top clamping beam needs to be locked in position tightly enough to trap the sheet, but its height needs to be set to take account of the thickness of the sheet, especially where the beam is clamped using a cam arrangement (Fig. 6-9). Screw-down clamping beams are fine for a single fold, but for repetition work the consistency of the clamping force is important because any significant variation will cause a difference in the curvature of the bend. Once set for the sheet thickness, cam-action clamps provide that consistency, and are quick to operate.

The thickness of the sheet and the mechanical properties of the material determine the minimum radius of bend because thicker material cannot be bent as sharply as thin sheets of the same material. The radius of the bend required, whether it is a minimum radius bend or a gentler bend with a larger radius, also affects the alignment of the front of the top clamp in relation to the hinged beam in its operating position (Fig. 6-10). The top clamp beam must be drawn back to accommodate the radius of the bend when the hinge beam has made the bend, otherwise the material may become trapped, and the bender will jam.

This is known as the *backset*, and it should be equal to the internal radius of the bend, plus the thickness of the sheet, or a distance equal to the external radius of the bend (which should amount to the same distance). This can be set approximately by using a ruler or a digital calliper, but a better way is to make a simple gauge strip of material with its thickness equal to the required distance, then use this as shown in Figs 6-11 and 6-12 to set the distance between the hinged beam and the top plate.

Fig. 6-13: Backset adjustment eccentric collar, with grubscrew lock.

Fig. 6-13 shows an eccentric collar on the rear of a box and pan folder, to adjust the backset. There is a similar collar at the other end of the folder, and both need to be adjusted to ensure the backset is equal all the way along the front edge of the fingers.

Using a gauge takes account of any clearance caused by the action of the hinge on the bending beam. Some care is required in setting this distance using a gauge strip, because of the thin front edge of the clamp bar, and it is important that the gauge strip is pushed fully down on to the fixed bed of the folder so that it sits between the front of the top clamp bar and the hinged beam.

If the clamping bar is lowered and locked by a cam, it will swing inwards slightly as it descends, so it is important that it is allowed to rest gently on a piece of sheet material of the same thickness as will be bent, as the backset is adjusted.

However, the top clamp bar may have a small radius on its nose (Fig. 6-14), and it is difficult to measure this, so unless that radius has been specified by the manufacturer, the final radius of bend may need to be set by experiment. On a small folder, the effect of the nose radius is likely to be small because the nose may not have been machined to a specific radius and it is usually possible to ignore the effect. In practice, the effect will vary from one make and model of folder to another, and may be greater on a larger or more expensive folder. Some folders are deliberately designed to have a large nose radius, especially where the deformation of the material must be carefully controlled – for example, during the manufacture of aircraft parts, where the finished part will be subject to mechanical test and inspection before certification.

On most small bending machines the backset must be set using a gauge strip, but a refinement would be a calibrated scale or calibrated backset adjuster fitted to the bender. This would allow the required backset to be adjusted by reference

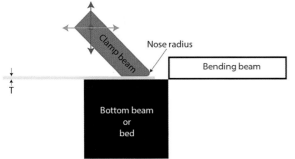

Fig. 6-14: The clamp beam may have a small nose radius.

to the scale, and would remove the need for gauge strips. This method would not remove the need to make some trial bends to estimate the effect of the nose radius. However, once the effect of the radius is known, adjusting the zero point on the scale would allow this to be permanently included in the backset.

ANGLE OF BEND

Swinging the bending beam upwards bends the sheet, and as long as the stress exceeds the elastic limit of the sheet, a permanent bend will occur. However, some of the deformation will be permanent, and some will not, and once the bending beam has been swung back down, the sheet will tend to spring back a little, so a sheet will normally need to be over-bent a little, bending it more than is required for its final position. Then, when the bending beam is swung out of the way, the sheet will spring back, leaving it permanently bent at the angle required.

The amount of spring-back will vary depending on the material, and can be found by a little experiment, or by bending, checking the angle, then bending a little further, until the required angle is reached. It is not a good idea to create a permanent bend of more than is required, then try to open out the bend a little, because that will cause a more complex deformation in the area around the bend, and it is likely to be difficult to remedy this without causing further deformation or damage, spoiling the workpiece.

For a given thickness and length, a material with low elasticity will have a reduced tendency to spring back, and will require little over-bending, while a material with higher elasticity will have a greater tendency to spring back and will require a greater degree of over-bending.

Material suppliers can normally provide a minimum bending factor (MBF), and a spring-back factor (SBF), which can be used to calculate the over-bend required to produce a particular angle of permanent bend in a sheet.

Note that the spring-back factor is sometimes denoted by the letter 'K'. This can be misleading, though, as K is also used to denote the ratio of movement of the neutral axis when calculating the flat pattern length around a bend. In this book, spring-back factor is denoted by SBF, and K is reserved for use when working out flat pattern distances (see later).

For example, Table 10 shows the typical minimum bending factors for aluminium sheet 1050A – H14 between 0.5mm and 12mm thick.

Table 10: Typical minimum bending factors for aluminium sheet 1050A – H14

Sheet thickness (mm)		Minimum bending factor	
From	To	90°	180°
0.5	1.5	0.5	1.0
1.51	3.0	1.0	1.0
3.01	6.0	1.5	1.5
6.01	12.0	2.5	---

Fig. 6-15 shows 90-degree and 180-degree bending angles in sheet. On a drawing, sheet can be bent back on itself without difficulty, but in the real world there is a limit imposed by the thickness of the sheet, and the difficulty of bending the sheet increases with sheet thickness. Bending aluminium sheet 1050A – H14 thicker than 6mm through 180

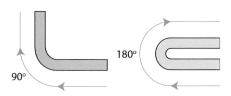

Fig. 6-15: 90-degree and 180-degree bending angles, with the larger angle showing the difficulty of bending thick sheet back on itself.

degrees is a practical impossibility without splitting the sheet, which explains why there is no entry for 6.01 to 12mm sheet at 180 degrees.

Using the simple bending guidelines:

Minimum bending radius =
thickness × minimum bending factor

The following calculations may be made:

For a sheet 1.0mm thick, and a 90° bend, the minimum bending factor is 0.5
minimum bending radius = 1.0×0.5 = 0.5mm
For a sheet 3.2mm thick, and a 90° bend, the minimum bending factor is 1.5
minimum bending radius = $3.2 \times 1.5 = 4.8$mm

Sheet metal has a 'grain' that lies along the direction in which the sheet was rolled during manufacture. Because 90-degree bends are so common, some material suppliers provide more detailed bending guidelines for this material (see Table 11) when the line of the bend is at right angles to the direction in which the sheet was rolled (Fig. 6-16). The direction of rolling is

Fig. 6-16: The 'grain' of a sheet may be visible: in this photograph it runs left to right.

normally along the length of the sheet, but there are exceptions to this.

Recognizing the practical limits on bending sheet, manufacturers may supply details of recommended minimum (internal) radius of bend for each thickness, as shown in Table 11.

The spring-back factor (SBF) is the ratio of the angle of the bend after spring-back to the angle through which the folder bent the sheet before releasing it to spring back (Fig. 6-17). Spring-back depends on the ratio of the bending radius (after spring-back) to the sheet thickness, as shown in Table 12.

For a 90-degree bend in a sheet 1.0mm thick, and a bending radius of 2mm:

bending radius : sheet thickness = 2:1 = 2

Table 11: Sheet thickness and minimum radius of bend for aluminium 1050A – H14

Sheet thickness (mm)									
0.5	1.0	1.5	2.0	3.0	4.0	5.0	6.0	8.0	10.0
Minimum inner radius of bend after spring-back (mm)									
0.3	0.5	1.0	1.5	2.5	3.8	5.5	7.5	13	18

Table 12: Spring-back factor for aluminium 1050A – H14

Ratio of bending radius to sheet thickness							
1	2	4	6	10	15	30	60
Spring-back factor (SBF)							
0.98	0.98	0.97	0.95	0.93	0.90	0.80	0.65

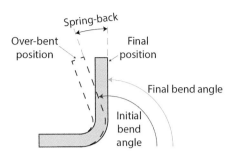

Fig. 6-17: The final bend angle depends on the amount of spring-back.

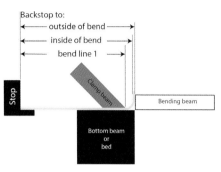

Fig. 6-18: Distances may be measured from a backstop.

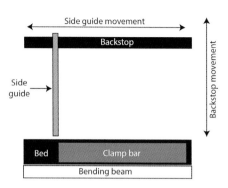

Fig. 6-19: A side guide locates the edge of a workpiece in relation to the fingers.

From the table, a ratio of 2 corresponds to a spring-back factor of 0.98; this means that when the sheet is bent to 90 degrees the permanent deformation will only be:

$0.98 \times 90° = 88.2°$

To find the angle of bend required to result in a permanent deformation of 90 degrees, divide the intended angle by the spring-back factor. So, for a 90-degree bend, this material needs to be bent through 90/0.98 = 91.8 degrees (approximately 92 degrees). Bending to 92 degrees will allow the sheet to spring back to 90 degrees when released from the folder.

Be aware, however, that this is a theoretical calculation, and although it is extremely useful, the real-world results may differ slightly. The best way to check is to do a test bend using the same material with the same thickness and the same bend radius as the final workpiece. However, it is better to know (by calculation) that a 2-degree over-bend is needed rather than a 10-degree over-bend.

On most small folders, the bend angle needs to be estimated by eye, but alternatives include a fixed scale at one end of the folder to allow angles to be judged accurately, or some simple gauges cut to specific angles to set the final position of the bending beam by comparison with the appropriate gauge.

For repetition work, three additions are useful on a folder. The first is a backstop (Fig. 6-18) so that the fold line is always at a known distance from the backstop and therefore from the front of the bed of the folder. The backstop can be set to measure distance as shown in Fig. 6-18, as follows:

- either from the stop to the front of the clamp beam, which would be the distance to the beginning of the inner bend line
- from the stop to a gauge strip of the same thickness as the material to be bent, set against the final position of the face of the bending beam, which would be the distance to the inside of the finished bend
- or from the stop to the final position of the face of the bending beam, which would be the distance to the outside of the finished bend

The second useful addition is a side guide. This is an arm that normally sits at right angles to the backstop and locates the edge of the material in relation to the fingers. It may also be adjustable so that it can slide along the rear guide to position the edge of the material accurately across the bed of the folder (Fig. 6-19).

The third useful addition is a stop for the fold angle so that the bending beam always folds the sheet to the same angle (Fig. 6-20).

Setting the backstop, the side guide and the angle stop will allow a considerable degree of accuracy of fold in a batch of identical work.

Fig. 6-20: An angle stop (the circular disc) helps ensure accurate folds during repeated operations.

An Example Bend

This example bend uses external measurements to fold a workpiece, as shown in Fig. 6-21:

Material: aluminium 1050A

Thickness: 1.2mm (so minimum bending factor = 0.5, and minimum internal radius after bending will be a little under 1.0, which is the MBR of material 1.5mm thick)

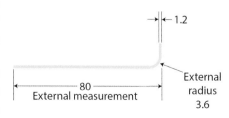

Material: aluminium 1050A
Thickness: 1.2mm
Angle of bend: 90°

Fig. 6-21: An example bend.

For the dimensions shown, the external bend radius is 3T, that is (3 × 1.2 = 3.6mm), so the internal bend radius is 2T, that is (2 × 1.2 = 2.4mm). The internal bend radius is needed because this is the smallest radius, and it must be not less than the MBR for the material.

Ratio of bending radius : sheet thickness = 2T : thickness = 2.4 : 1.2 = 2

Spring-back factor = 0.98

For a 90-degree bend, the required bending angle is 90°/0.98 = 92°

To set up for this bend (Fig. 6-22):

- Adjust the backset to the external bend radius = 3.6mm
- Set the backstop to the distance from the rear edge to the face of the raised bending beam = 80mm
- Set the bend stop to 92 degrees

FLAT PATTERN LENGTH

Every bend has a radius, and a sheet deforms as it bends, so these factors need to be taken into account in working out the length of sheet required to make a particular bend and end up with the expected lengths on either side of the bend. Fig. 6-23 shows a finished bend in which there are two flat portions of sheet (with lengths denoted as L1 and L2) and a bend. It is the length of the bend that varies depending on the angle of bend, the material and the bending method, and although it is not possible to get an exact figure for the length of that bend simply by calculation, it is possible to get close enough for practical purposes, and then adjust that figure in the light of practical results and experience.

When a sheet is bent, the material on the inside of the bend will be compressed, while the material on the outside of the bend will be stretched. The neutral axis (NA) is the imaginary line along the axis of the sheet on which the material will be neither compressed nor stretched. In theory, the NA will run through the middle of the cross-section, half the thickness (0.5 × T) away from the face of the sheet, but in practice it may be closer to the inner side of the bend. In fact the tighter the bend, the more compression on the inner side of the bend, and the more the NA moves towards the inner side. To cope with this, we can make allowances by using the ratio of the distance from the inner face of the material to the whole thickness of the material (Fig. 6-24). This is known as the K factor, and it is used in calculations associated with bending (see below).

Material suppliers may be able to state the K values for their materials, but in practice,

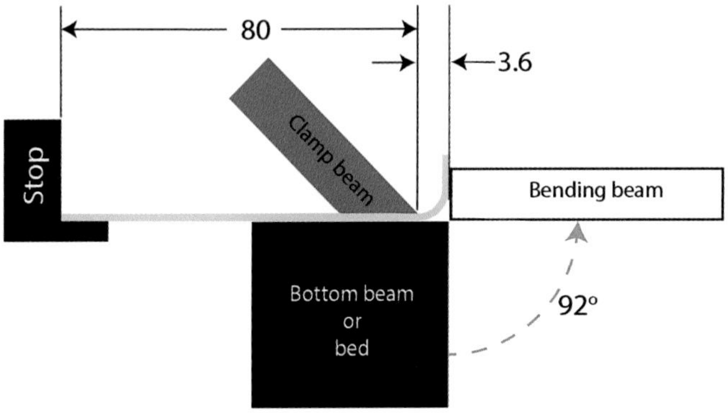

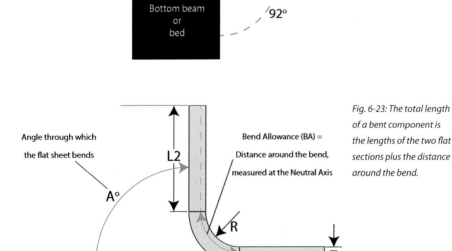

Fig. 6-22: Setting up for the example bend.

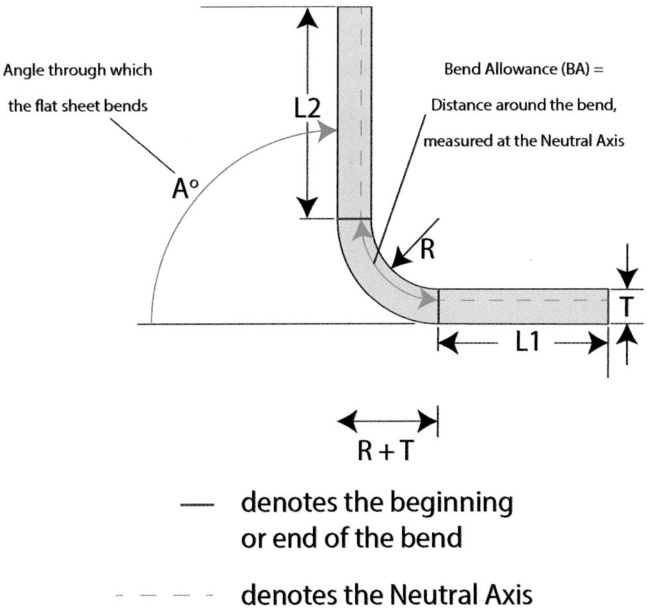

Fig. 6-23: The total length of a bent component is the lengths of the two flat sections plus the distance around the bend.

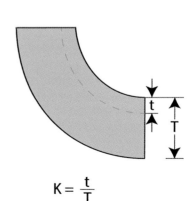

$$K = \frac{t}{T}$$

Fig. 6-24: Material thickness (T), distance from the inner face to the neutral axis (t), and the K factor t/T.

it is often difficult to obtain an accurate theoretical value, and K may be best determined for a particular material and bending method by carrying out some test bends. Alternatively, use these values, all of which are *approximate* and based on the simple fractions ¼, ⅓ and ½, as shown in Table 13.

Table 13: Rule-of-thumb approximate K values for different radii of bend, where precise values are not available

Bend radius compared to thickness	K factor
Bend radius less than 1T (i.e. 1 thickness)	1/4 or 0.25
Bend radius between 1T and 2T	1/3 or 0.33
Bend radius greater than 2T	1/2 or 0.5

Although Table 13 provides an approximation for K, it is only a starting point, and the actual value should be determined by making some test bends (*see* later). Once K is known, the flat pattern length can be calculated by one of two methods:

◆ Calculate a *bend allowance* (BA), which is the length around the bend, at the NA. To find the flat pattern length, add the flat lengths before and after the bend, and the bend allowance (*see* Fig. 6-23).
◆ Calculate a *bend deduction* (BD) as well as a bend allowance. To find the flat pattern length, add the lengths measured to the tangent points of the outside of the bend (Figs 6-25 and 6-26) or to the apex point (Figs 6-25 and 6-27), then subtract the bend deduction. The BD takes account of the BA, but adjusts the length to take account of the position of the point from which the measurements are taken.

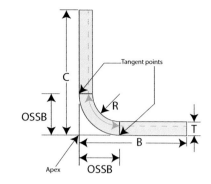

— denotes the beginning or end of the bend
---- denotes the Neutral Axis

Fig. 6-25: For a right-angle bend, the distances to the tangent points of the outside of the bends are the same as the distances to the apex.

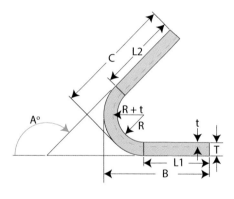

L1 and L2 play no part in this method

Fig. 6-26: Distances to the tangent points on a reflex bend.

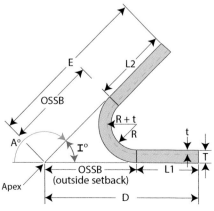

Fig. 6-27: The outside setback lengths on a reflex bend.

Calculating Bend Allowance

Bend allowance (BA) is simply the length of material (in the flat pattern) required to make a particular bend. We could use geometry to calculate the BA and, in a perfect world, that would be an exact match for the physical finished workpiece. In practice, we use a K factor to modify the calculated length around the bend so that it more accurately reflects what happens in real life.

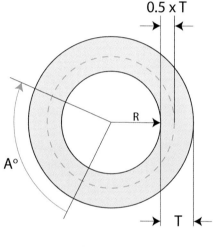

---- denotes the Neutral Axis

Fig. 6-28: The circumference of the bent material is calculated using the radius of the curve at the neutral axis.

If the NA runs along the centre of the material, Fig. 6-28 shows that the circumference of a complete circle would be π × diameter, or π × 2 × (R + 0.5 × T). For less than a full circle, use proportion based on the bend angle, so that the length of the section BA would be:

$$(A/360) \times 2 \times \pi \times (R + 0.5 \times T)$$

where A is the angle of bend, in degrees.

Where the NA moves because of the bending action, modify T using K, so that:

$$BA = (A/360) \times 2 \times \pi \times (R + K \times T)$$

For a 90-degree bend, this can be simplified:

BA = (90/360) × 2 × π × (R + K × T)
BA = ½ × π × (R + K × T)

The flat pattern length (FPL) is then found by adding the BA to the sum of the two flat lengths, before and after the bend, L1 and L2 (see Fig. 6-23):

FPL = L1 + L2 + BA

Calculating Bend Deduction

Bend deduction (BD) recognizes that it is often easier to measure the finished sizes of a bent workpiece by referring to the tangent points where the outside faces turn (see Fig. 6-26) or to the apex of the bend (see Fig. 6-27). In the case of a 90-degree bend, these are the same sizes (see Fig. 6-25).

In either case, BD incorporates BA so that the flat pattern is found by adding the lengths and subtracting the BD.

Measuring to the tangent points (Figs 6-26 and 6-25), add the two lengths B and C:

L1 = B − (R + T)
L2 = C − (R + T)
FPL = L1 + L2 + BA
    = B − (R + T) + C − (R + T) + BA
    = B + C − 2 × (R − T) + BA

FPL = B + C − BD, where BD = 2 × (R + T) − BA

If the measurements are taken from the apex (see Figs 6-27 and 6-25), calculate the included angle of bend, I degree, where I degree = 180 − A degrees (for a right-angle bend this will still be 90 degrees because 180 − 90 = 90).

Calculate the outside setback (OSSB), which is the distance from the apex to the beginning or end of the bend (and both distances are equal), as:

OSSB = (R + T) / Tan (I/2)

The bend deduction is twice the OSSB minus the BA, so:

BD = 2 × OSSB − BA
FPL = D + E − BD

Calculating K from a Test Bend

Take a piece of the material you will use for your job. Measure the thickness accurately (say, 1mm), and use a known length (say, 100mm long). Mark a fold line at 40mm from one end, set the folder to produce a radius of bend of twice the material thickness (2mm), then make a 90-degree bend.

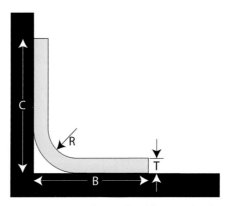

Fig. 6-29: Measuring distances B and C on a test bend.

Set the bent material on a flat surface, against a vertical edge (perhaps the upstanding face of an angle plate) as shown in Fig. 6-29, and measure the two distances B and C:

BD = B + C − total length
OSSB = (R + T)/tan 45 degrees = R + T
BA = 2 × OSSB − BD

$$K = \frac{\left(\frac{2 \times BA}{\pi}\right) - R}{T}$$

Make a careful note of the result. For improved accuracy, you could make more than one 90-degree bend, and average the results.

If you are using a CAD program to draw your sheet metal parts, and the flat shapes (developments), the program may allow you to use the value of K you have found using your test bends. That way, you should get more accurate results for that material, folded in that way, using that folding method.

BENDING THICKER SHEET

The thicker the sheet, the more force is required to bend it, so for anything over approximately 1.2mm mild steel, the folder needs to be of heavier construction. Longer folders suffer from the fact that a longer length of bend will require more force, and the longer the bending beam, clamp beam and bed, the greater the deflection under load. This means longer folders need to be of heavier construction.

The practical result is that the cost of heavier folders increases. Because of that, longer folders are often designed to accommodate thinner sheet. Folders designed for thicker material, such as 6mm sheet, may have a limited bending length. They may also have a significant nose radius on the clamp bar or the fingers, because the properties of the material will dictate a significant minimum bend radius at that thickness.

THE PRESS BRAKE

Press brakes (Fig. 6-30) press sheet material between an upper finger called a punch and a lower die, to bend it. Both punch and die are normally V-shaped (Fig 6-31). As the punch forces the sheet into the die, the sheet bends. The distance the punch descends into the die can be controlled and affects the radius of the bend. The maximum travel of the punch is to the point where the sheet fills the space between punch and die.

Press brakes are versatile, and common in larger sheet metal workshops, but the presses and the tooling are expensive. Because of the forces involved in bending what are often long lengths, press brakes are normally hydraulically operated. However,

Bending Sheet Metal • 71

Fig. 6-30: A modern press brake: the TRUMPF TruBend 1530. TRUMPF GROUP

Fig. 6-32: A punch and die set that can be used on the bed of a fly press or hydraulic press.

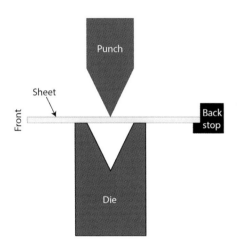

Fig. 6-31: Press brake punch and die.

in the same way as their larger industrial equivalents.

One limitation of the press brake is that it cannot bend more than 90 degrees except by using special tooling and procedures. The smaller manual machines are limited to angles of up to 90 degrees. For angles less than 90 degrees, the bend radius depends on the angle of the bend (Fig. 6-34), because in air bending the radius is not determined by a die.

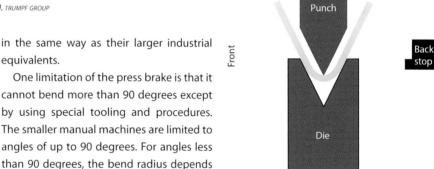

Fig. 6-34: The bend radius depends on the angle of bend: the smaller the included angle, the smaller the radius of the bend.

the press-brake action of punch and die can be used over shorter lengths in the popular hydraulic bench and floor presses often found in garages and workshops.

A separate press-brake unit can be fitted to the bed of the press, and operated via the press (Fig. 6-32). A similar tool is available for fly presses and for small arbor presses. The popular three-in-one, or Formit, machines (Fig. 6-33) incorporate a manually operated press brake. Although the punches are of a simplified shape, they nevertheless work

Fig. 6-33 The WARCO Formit 3-in-1 shear, press brake, and slip roll machine. WARREN MACHINE TOOLS LTD

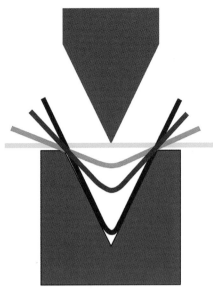

Fig. 6-35: In air bending, the angle of bend and the radius of the bend depend on the height to which the punch is lowered.

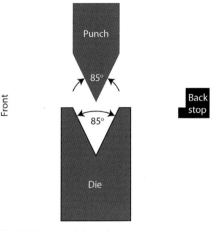

Fig. 6-36: Punch and die angles.

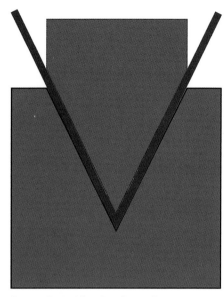

Fig. 6-38: During V bending, the punch presses material along the full depth of the die sides.

AIR BENDING

In a press brake, the angle of bend and the radius of the bend depend on the height to which the punch is lowered (Fig. 6-35), and the maximum angle of bend is set by the included angle of the die. This technique is known as air bending, perhaps because there is air underneath the bend line as the sheet is bent. Because of the need for over-bending to compensate for spring-back, punches designed to create 90-degree bends commonly have a nose angle of 85 degrees, and dies have an included angle of 85 degrees (Fig. 6-36). Lowering the die increases the angle of bend, progressively, until the punch holds the sheet against the sides of the die.

The angle and the radius of bend will vary from a shallow angle and large radius where the punch has just begun descending, to a more predictable 90-degree angle and a small radius when the punch reaches full depth. Bending to a specific angle less than 90 degrees requires accurate control of the depth of the punch. On a small manual press brake, an angle of bend less than 90 degrees is best judged against a template which takes account of spring-back for the material being bent. Templates are best developed by making test bends.

Like the box and pan folder, most press brakes have segmented punches to allow previously bent sections to swing into the gaps at the sides of the punches. Many also have segmented lower dies, which allow unbent material to swing clear on both sides

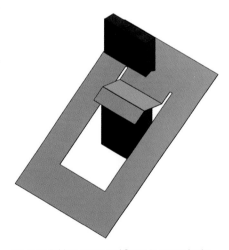

Fig. 6-37: Folding an internal flange in a press brake causes the sheet to fold down around the sides of the die.

of the punch and die. That allows internal flanges to be bent (Fig. 6-37).

V BENDING

In V bending, the V shape of the punch matches the V shape of the die so that at maximum travel the punch presses material along the full depth of the die sides (Fig. 6-38). This is a commonly used bending method, both on small manual benders and on large industrial press brakes.

BOTTOMING

Bottoming occurs when the punch compresses the material against the *sides* of the die at the end of its stroke (Fig. 6-39). At that point, the bend radius and the spring-back will be at their minimum. The nose radius of the punch will be greater than the radius of the inside of the bottom of the die, so contact is between the sides of the punch, the material, and the sides of the die. The nose of the punch does not form the bend at this point.

Bending Sheet Metal • 73

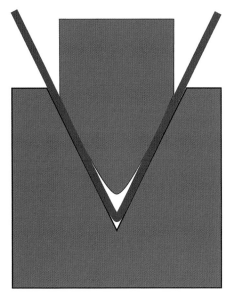

Fig. 6-39: Bottoming occurs when the punch compresses the material against the sides of the die at the end of its stroke.

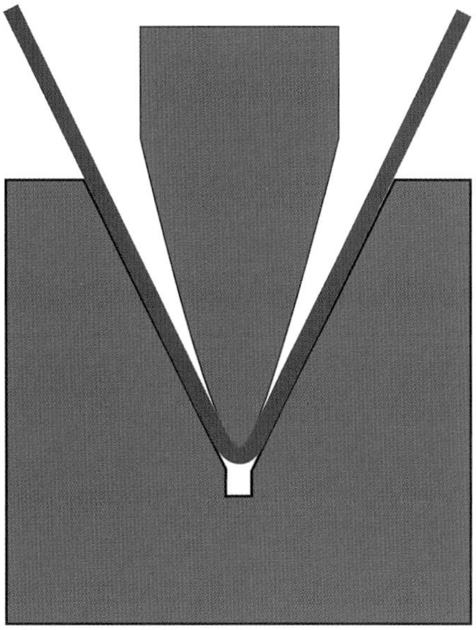

Fig. 6-40: During coining, the punch nose presses the material against the bottom of the die.

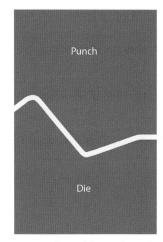

Fig. 6-42: Punches and dies can be used to make a wide variety of bends during multi-bending operations.

COINING

During coining, the nose of the punch forms the final bend by pressing the material against the bottom of the die (Fig. 6-40). The punch nose radius forms the inner radius of the bend by compressing the sheet at that point.

U BENDING

Making the end of the punch flat, with only a small chamfer along each corner, allows it to drag the material down between a die on each side (Fig. 6-41). The punch effectively performs two bending operations simultaneously, one at each corner of the punch.

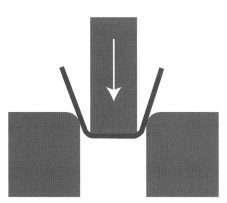

Fig. 6-41: In U bending, the punch drags the material down between a die on each side.

The dies are likely to require a radius at the corner contacting the material, and the surface finish will be reflected in the outside surface of the material.

Industrially, the punch often presses down on to a central pad, which may only support part of the underside of the material. The size and the resilience or springiness of the pad can be used to control the spring-back and angle of the final bends.

MULTI BENDING

Appropriately shaped punches and dies can be used to make a wide variety of bends simultaneously, as shown in Fig. 6-42. The design of punches and dies is a specialist skill, but a well-designed profile can simplify combined folds and speed mass production.

7   Rolling, Beading, Flanging and Wiring

Apart from folding, sheet material can be curved by rolling, and can have its shape changed by a variety of operations involving rotary dies or form wheels.

ROLLING

A flat sheet can be curved by passing it through a set of three rolls (Fig. 7-1). A typical set-up involves three rolls, with the sheet being passed through a pair of rollers stacked one above the other. These grip and push the sheet towards a third roller at the rear. The height of that third roller can be adjusted (Fig. 7-2), and it forces the sheet upwards, curving it in the process. By adjusting the height of the third roller, the tightness of the curve can be controlled to produce anything from a gentle curve to a tight cylinder.

On a set of 'slip' rolls the top front roller can be removed or at least freed at one end to allow tightly curved work to be removed by sliding it sideways off the roller (Fig. 7-3). Some rolls also have semi-circular grooves at the ends, to allow wire to be rolled and curved in the same way as sheet (visible at the right-hand end of the rollers in Fig. 7-3).

One of the disadvantages of this arrangement of three rolls is that they tend to leave an initial section of the sheet flatter than the remainder, and that may need further attention after removal from the rolls. The

OPPOSITE PAGE:
Railway signal lamp from the Great Northern Railway. Made in large numbers, these lamps demonstrate many of the sheet metal and tinsmith's skills.

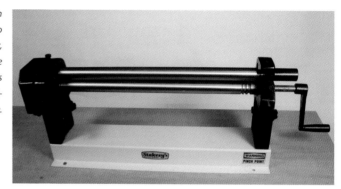

Fig. 7-1: Geared rolls with three rollers. The front two rolls are geared together, and have an adjustable gap. The rear roll is free-running and height-adjustable.

Fig. 7-2: The rear roll has a height adjuster at each end (the knob on the right of the photo). The gap between the two front rolls can be altered, to vary the grip (the knob on the left side of the rolls).

Fig. 7-3: Slip rolls allow the top roll to be swung free, to allow tightly curved work to be removed.

effect can often be remedied by rolling a greater length than is required, then trimming the flatter section from the sheet after removal from the rolls, or by passing the sheet through the rolls, turning it lengthways, then passing it through again with the initial flat section going through the rolls last.

Setting the rolls parallel will result in a curve which is uniform across the sheet. Taking several passes, raising the height of the rear roller each time, will eventually result in a cylinder.

Setting the rear roll with one end higher than the other will produce a conical shape with one end more tightly rolled than the other (Fig. 7-4). It will not be possible to close the cone fully at the top, though, because the sheet cannot be rolled tighter than the diameter of the roller, so this will be a truncated cone.

Where a component is to have a curved section alongside flat sections, it may be possible to roll just that part of the component, particularly if the bend is at one end. In other cases, a combination bending roller may be used. This has a bending action, but instead of the bending beam being flat it takes the form of a roller that 'wipes' the sheet around a circular former. Fig. 7-5 illustrates the action. This is a useful machine, but each radius of bend requires a specific former. In addition, the formers need to be chosen to give the required bend after spring-back, and that may vary from one thickness of material to another.

A curved section can also be produced using a tube or rounded punch as a former, pressing the sheet between two guides to create a bend within the sheet (Fig. 7-6). The guides may themselves be rollers, or they may have straight sides. This operation can be done in a press brake, if suitable fingers and formers are available. Short sections can also be produced in a fly press, with similar tooling (Fig. 7-7).

Fig. 7-4: The minimum diameter of a truncated cone is dictated by the diameter of the roller. Use slip rolls, where the top roller can be removed, to release the cone.

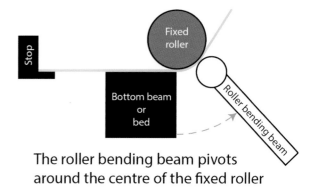

Fig. 7-5: A combination bending roller uses a wiping action to fold the material around a former.

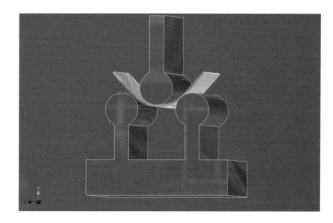

Fig. 7-6: Using a curved punch to press a sheet between two rollers curves the sheet.

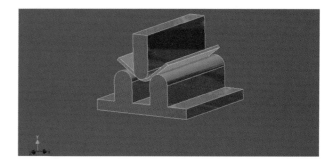

Fig. 7-7: A curved punch-bending tool can be used in a fly press.

Fig. 7-8: A flat sheet can be put through a series of rollers, which bend it into a shape.

Fig. 7-9: Manually operated jenny or bead roller.

Rollforming

A rollforming machine uses several sets of rollers that act in pairs to gently form a flat strip into a more complex section, in stages, as the strip passes through the rolls. Fig. 7-8 shows the action. Because there are several sets of rollers, and bending takes place in stages, this is a relatively low-stress operation compared to forming the bends in one operation.

Lockformers are rollforming machines that produce a specific profile, usually on the edge of a sheet, as the sheet passes through the rollers. The profile is designed to interlock with an adjacent sheet to produce a join. For more details, *see* Chapter 9.

The Jenny

Known by various names, the jenny, swage or bead roller (Fig. 7-9) uses pairs of die rollers acting on either side of a sheet component to create a bend, bead, flange, fold or flare. Hand- or power-operated, this is a versatile machine capable of a range of operations. The traditional jenny has a 150 to 175mm throat, but larger versions of a different design are available with much larger throats of up to 1m.

Bending takes place between stepped dies designed to act in pairs to change the shape of a section of a sheet. Place the sheet between the rollers, adjust the pressure of the rollers on the sheet, then turn the handle to rotate the rollers, which move the sheet through while impressing a shape on the sheet.

A wide range of dies is available, depending on the make and model of jenny, so this machine is capable of a wide range of tasks.

The jenny is often fitted with a backstop to guide the material and set the distance from the rear edge to the feature being produced (Fig. 7-10).

Fig. 7-10: Adjustable backstop (shown behind the lower roller) carrying two flat pads and which can slide along the lower shaft.

The lower roller rotates about a fixed axis, and the shafts are geared together, but the upper roller is hinged at the rear, so there are adjustments on the top roller for:

- adjusting the top die forwards or backwards in relation to the lower die, to allow the dies to be aligned
- varying the pressure on the sheet
- bringing the roller and shaft parallel to the lower roller and shaft

BEADING

Beading involves pressing a semi-circular groove into the sheet (Fig. 7-11), and this is done with one roller shaped as a bead and the other as a matching die (Fig. 7-12). The groove can be straight, or it can follow a curve by manually guiding the sheet as it passes through the rolls.

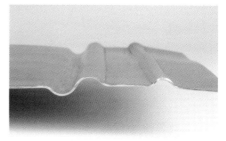

Fig. 7-11: A plain bead on the right, and an ogee bead on the left.

Rolls dies of different radii are available, to give different sizes of bead.

Rolls and dies can carry other shapes for this operation, so that grooves with different radii can be rolled, or a square-shouldered

78 • Rolling, Beading, Flanging and Wiring

Fig. 7-12: Beading dies.

Fig. 7-13: Ogee dies.

Fig. 7-14: A joggle on the left, and a flange or burr on the right.

FLANGING

A flange or upstand (Fig. 7-14) can be produced using two rollers which pinch the sheet against a small step (Fig. 7-15). The upper roll is chamfered to allow the operator to encourage the sheet upwards as it passes through the rollers. Working in stages, passing the sheet through the rolls more than once and adjusting the pressure each time, a small lip is produced at right angles to the sheet. This operation is sometimes called 'burring', as the width of the flange that can be produced is limited in size as the upturning edge of the sheet collides with the top roller or shaft.

Fig. 7-15: Flanging dies.

Joggling

A variation of flanging/burring is to use a pair of dies that create a step or joggle (see Fig. 7-14). The dies create a flange and reverse flange simultaneously, by limiting the distance available for the upstanding flange before the material is forced into a reverse flange by the upper roller (Fig. 7-16).

Fig. 7-16: Joggling dies.

Crimping

Crimping dies (Fig. 7-17) produce a wavy edge on a sheet (Fig. 7-18). This has the effect of shrinking the edge slightly, but it is

groove or bead can be produced. More complex shapes include ogee beads that have a bead and a reverse bead sitting next to one another (see Fig. 7-11). The two beads forming the ogee are produced simultaneously, using a pair of matching ogee dies (Fig. 7-13).

Fig. 7-17: Crimping wheels.

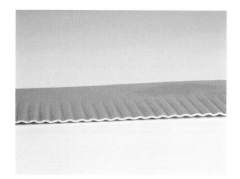

Fig. 7-18: A crimped edge.

Fig. 7-19: Wired edge.

Fig. 7-20: Wired edge: fold over a radiused former.

Fig. 7-21: Wired edge: close the flange on the former.

also useful for producing a slip joint in two identical rolled tubes. Crimping the end of one tube reduces its diameter, allowing it to slip inside the other tube. Because the tubes now only touch along the tops of the wavy crimp, friction is much less than when trying to push one tightly fitting tube into the end of another. Adjusting the pressure between the dies allows good control of the amount of shrinkage and the tightness of fit.

WIRING AN EDGE

A wired edge involves forming a rolled-over curl on the edge of a sheet, and a wire trapped inside the curl (Fig. 7-19). This can be done by hand, with careful hammer work, by using a folder followed by hand-work, or by using the jenny combined with some handwork.

As a general rule of thumb, the fold line, or the beginning of the curve in a sheet to make a wired edge, is between two and a half and three times the diameter of the wire. For a typical 3.2mm wire, mark the fold line 8mm from the edge for a roll that closes very tightly and uniformly (usually requiring a jenny or similar machine closure), 9mm for a good fit, and up to 10mm for a very easy fit, including use as a hinge where the rolled edge must be able to rotate around the wire.

A hand-wired edge requires a former around which to make the initial curved flange in the sheet. Ideally, that should have a radiused edge of the same diameter as the wire. Working upside down initially, with the former face down and the sheet uppermost, use a mallet or a hammer with a non-metallic face to work along the edge, gradually turning it over the former until it has turned at least 90 degrees, and preferably a little more (Fig. 7-20).

Flip the former and sheet over, and continue folding to make a channel that looks like a U shape lying on its side (Fig. 7-21). Remove the former and insert the wire (Fig. 7-22), holding it tightly in place, and

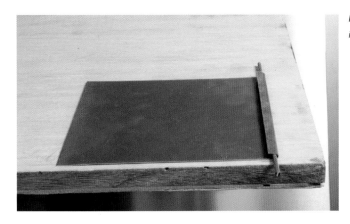

Fig. 7-22: Wired edge: insert the wire.

Fig. 7-25: Small flat on the nose of a hardwood striker.

Fig. 7-23: Wired edge: use a striker to close the flange around the wire.

on the final pass the sheet will rub against the upper roller, and it is important that the nut securing that roller does not protrude and scuff the sheet. At the workbench, put the wire into the groove and use a mallet or a cross-pein hammer to fold the edge of the sheet over the wire.

It is not necessary to completely tighten the fold at this stage, although the closer the better, up to the point where the mallet cannot strike the edge without also striking the main non-wired part of the sheet. Use a continue to fold the edge as far over the wire as possible, using the mallet or a planishing hammer. Stop before the adjacent face of the sheet is marked.

Butt the wired edge against a firm support and use a wedge-shaped striker (Fig. 7-23) to make the final closure (Fig. 7-24). The striker may be a hard plastic or hardwood block with a tapered nosepiece (Fig. 7-25), with a tiny flat at the end.

The purpose of the striker is to close the curl, but not to lever the wired edge off the sheet. Make the striking face of the striker as broad as possible, to spread the impact and avoid marking the surface of the curl.

The initial curl can be produced with a jenny by using a pair of beading dies, tilting the sheet upwards as it is rolled, to produce the J-shaped edge to the sheet (Fig. 7-26) – but note that a larger upstand will result in a

Fig. 7-24: Finished wired edge.

bead between two flat areas, while a smaller upstand will be difficult to produce without distortion. Scribe a line 8mm from the edge of the sheet, then position the scribed line directly under the top roller.

Run the sheet through perhaps four times, increasing the pressure and tilting upwards a little more at each pass. Make sure to follow the same path each time. Note that

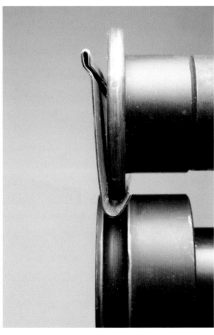

Fig. 7-26: Using beading dies to form a J-shaped edge profile, for wiring an edge.

piece of scrap plate, properly deburred, as a shield to protect the main sheet.

Put a pair of wiring dies into the jenny, to close the edge of the sheet tightly around the wire; or complete the closure by hand, using a striker. This can also be done by replacing the top roller as shown in Fig. 7-26 by a plain roller, and using the bottom die to close the edge tightly. Hold the flat section of the sheet against the plain roller, with the wired edge inside the groove in the lower die.

Producing a rectangular shape that is wired on all edges can be done in stages. The wire should ideally be continuous, bending around the corners and with the ends meeting somewhere inside one edge. The method is the same, but the wire needs to be bent to shape at the corner positions (Fig. 7-27):

- Bend the wire to the rectangular shape
- Produce the J fold in each of the four sides, by hand or by using the jenny
- Trap the wire manually
- Complete the folds manually or in the jenny, but don't try to run the jenny round the corners as they are too tight. Conventionally, the wire is left uncovered anyway, because the developed shape of the sheet is simpler that way

It is also possible to wire a cylindrical edge, and the best way to produce the initial J curl is by using a jenny. Large-diameter cylinders pose few real problems, but the smaller the diameter, the more the upper die will scuff the adjacent parts of the curved edge, because that is curling in towards the edges of the die. Roll the wire into a circle before dropping it into the curl, then close the curl, initially by hand, then by using the jenny. There is an example in the project 'Spigot for a workshop dust extractor'.

Producing the J upturn on a cylinder involves stretching the edge, and closing the curl involves shrinking it, but there is no convenient way of doing this as a separate operation, so firm action in the jenny is the best way to persuade the edge to turn. The edge could be stretched with a hammer (see later section), but on such a narrow section, this would require careful work over a cylindrical dolly.

Shrinking will be required over a very small distance and in an awkward position as the edge is finally closed, so this can be ignored. So, stretching and shrinking are possible, but unnecessary. The larger the diameter of the cylinder, the smaller the percentage of stretching and shrinking required – so in practice, these operations can be ignored.

Some die manufacturers have produced dies designed for wiring edges in specific locations, such as a particular make and model of car wing, and those dies are shaped to produce a final beading action to make the wired edge stand proud of the panel on the outside.

Fig. 7-27: Edges can be wired round a corner, but the wire is very often left exposed as this simplifies the developed shape.

8 Joining

Some sheet metal objects are complete in themselves, while others serve along with other components as part of an assembly. The methods used to fix sheet components to other items, whether loosely or tightly, permanently or to allow disassembly, depend on the nature of the components and the mating parts. Sheet parts may be fastened to other sheet components, or they may be secured to a substantial, thick part of an assembly. Design choices mean that the mechanical properties required of a finished assembly – the characteristics of bolts, rivets, glue or weld, and the aesthetics of the finished article – will all influence the choice of assembly techniques and any fixings or fasteners used.

FIXING TO THICK COMPONENTS

Thick components (approximately 3mm and over) have enough depth to be drilled and threaded to allow a machine screw to pull a thin sheet tightly against a thicker part. As a rule of thumb, there should be a minimum of three threads in the thicker component to allow enough grip for a secure fastening. However, this will be affected by the material being threaded, and three threads in a soft plastic will provide a lot less security than three threads in a steel plate.

OPPOSITE PAGE:
Over 1,000 hand-set rivets. A Gauge 1 model of the tender of an A1 locomotive built by Dave Parker at Buxton Model Works. Every rivet hole was marked out with dividers, scriber and rule.
DAVE PARKER

FIXINGS AND FASTENINGS

Fixings and fasteners join things together. Fixings are normally designed to join things permanently, while fasteners are mechanical devices that not only join things together, but normally allow the joint to be disassembled. Rivets are fixings, for example, while nuts and bolts are fasteners. Although it is sometimes possible to remove rivets by drilling through them, they are designed to grip and prevent the joint from coming apart. Nuts and bolts, on the other hand, may join things securely and very firmly, but they are designed to be unscrewed to release the parts, using tools intended for that job (such as spanners).

Table 14 suggests minimum thicknesses for common thread sizes in sheet material. The pitch of thread being used will depend on the mechanical properties of the materials used for the fastener and the threaded component, with softer materials such as aluminium or plastic requiring a thread with a coarser pitch because that tends to provide a stronger threadform.

The height of a threadform, measured from the root to the crest, resists the tendency for a screw to pull out of a threaded hole (Fig. 8-1). In practice, the theoretical thread height is reduced because the percentage engagement of a screw is normally less than 100 per cent, to allow a working clearance and a manufacturing tolerance. However, this means a smaller effective height, resisting pull-out (Fig. 8-2), as well as a smaller effective width, resisting shear. This all brings the thread closer to stripping. The greater the percentage engagement, the larger the number of threads, and the longer the screw, the more secure the grip.

Table 14: Maximum thread pitch in common sheet thicknesses

Sheet thickness (mm)	Maximum thread pitch	Maximum thread size
6.0	2	M16 × 2
5.0	1.5	M10 × 1.5
4.0	1.25	M8 × 1.25
3.0	1	M6 × 1
2.5	0.8	M5 × 0.8
2.0	0.6	M3.5 × 0.6
1.6	0.5	M3 × 0.5
1.5	0.5	M3 × 0.5
1.25	0.4	M2 × 0.4
1.2	0.4	M2 × 0.4
1.0	0.3	M1.4 × 0.3*
0.9	0.3	M1.4 × 0.3*
0.7	0.2	M0.8 × 0.2**
0.6	0.2	M0.8 × 0.2**
0.5	0.16	-
0.4	0.13	-

* denotes available but non-standard thread series
** denotes non-standard threads not readily available

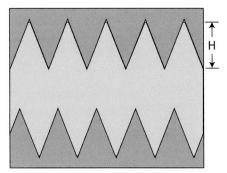

Thread height H at full engagement

Fig. 8-1: The height of a threadform is measured from crest to root.

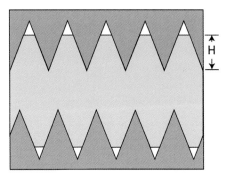

Thread height H at 75% engagement

Fig. 8-2: Reducing the depth of engagement reduces the effective height of the threadform.

In soft materials such as plastics, the low strength of the material means that threads are often formed without a separate tapping operation. Thus a fastener may be screwed directly into the material with only a small pilot hole to guide its initial positioning. This allows much closer to 100 per cent thread engagement, and may improve the strength of the material around the thread by compressing the material locally because it is pushed aside rather than being removed by the cutting action of a tap. To enhance this compression, fasteners to be screwed into soft material may have an increased root diameter (Fig. 8-3), as often seen on self-assembly furniture where a sheet metal component is to be secured to low-strength wood particle board.

Fig. 8-3: Threads for use in soft materials may have a coarse thread and a substantial core diameter.

Tapping

Producing a threaded hole in a thicker component conventionally involves drilling a pilot 'tapping size' hole, then using a tap to cut a thread around the inside of the hole. The diameter of the hole depends on the diameter and pitch of the thread as well as the intended percentage thread engagement. There are agreed standards for 'normal' engagement of standard threads, as shown in Table 15, which lists tapping drill sizes for common threads.

Tapping can be carried out by hand or with the aid of a machine such as a drill, milling machine or lathe. Tapping by hand involves using a tap designed for hand tapping, and a tap wrench which is used to grip the end of the tap and allow it to be turned in the hole to form the thread (Fig. 8-4). For each size of thread there are three hand taps in a full set: a *taper* tap with a taper ground along the first six threads, a *second* tap with a shorter taper at the front, and a *bottoming* or *plug* tap which has a taper only on the first thread. Note that the USA names these taps in a different way.

Making sure the tap starts upright, push the taper tap into the hole and twist it clockwise for a conventional right-handed thread, or anticlockwise for a left-handed thread. Once the tap has begun cutting a thread, repeatedly rotate it one turn clockwise then a quarter turn anticlockwise: this

Table 15: Tapping drill sizes for common threads

Thread diameter (mm)	Pitch	Tapping drill
0.8	0.2	0.6
1.4	0.3	1.1
1.6	0.35	1.25
1.8	0.35	1.45
2	0.4	1.6
2.2	0.45	1.75
2.5	0.45	2.05
3*	0.35	2.65
3	0.5	2.5
3.5*	0.35	3.15
3.5	0.6	2.9
4*	0.5	3.5
4	0.7	3.3
5*	0.5	4.5
5	0.8	4.2
6*	0.75	5.2
6	1	5
8*	0.75	7.2
8	1	7
8	1.25	6.8
10*	1	9
10	1.5	8.5
12*	1	11
12*	1.25	10.8
12	1.75	10.2
16*	1	15
16	2	14

* denotes metric fine series thread

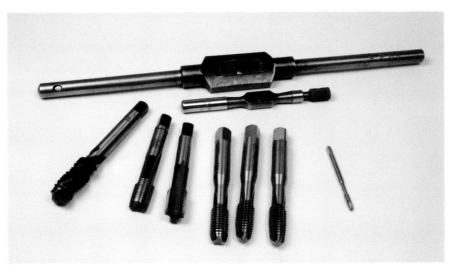

Fig. 8-4: Left to right, bottom row: spiral machine tap, two geometric (progressive) taps, a set of three hand taps (bottoming, second and taper), and a small partially fluted tap. Above the taps are tap wrenches.

Threading a punched 'tube'

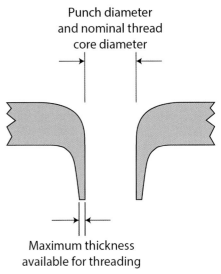

Fig. 8-5: Piercing a hole in a sheet using a punch allows a thread to be cut.

cuts a turn of the thread, then breaks any chips cut from the wall of the thread, so they can fall out of the hole or be accommodated in the flutes of the tap, to prevent the tap jamming. Carry on until the tap has reached the bottom of the hole or has passed right through the workpiece.

To finish threading to the bottom of a deep hole, use the second tap, followed by the bottoming tap. It may be necessary to wind the tap out of the hole every few turns if the hole is blind or the material clogs the flutes. In soft material or with a fine threadform it may be possible to omit the second tap.

Threading Thin Sheet

It is difficult to ensure enough threads inside a hole in thin sheet, even with a fine pitch thread. Fine pitch threads have a relatively small root-to-crest distance, and as a result have a comparatively low resistance to stripping. This makes a conventional threading operation of little use in thin sheet. So instead of punching a hole at the standard tapping diameter, a pointed piercing punch can be driven through, stretching the material at the sides of the hole to form a short tube, thereby increasing the effective thickness available for threading (Fig. 8-5).

Tapping then produces many more threads on the inside of the hole, and this allows a better combination of thread height and number of threads. However, the material forming the sides of the hole will not be thicker than the sheet, and this must accommodate the height of the threadform while leaving enough material between the crest of the thread and the outside of the tube to ensure the strength of the tube, as well as resisting stripping.

Conventional tapping produces a thread by removing material, weakening the remaining tube. Strength can be improved by using a fluteless tap or a roll tap, which forms a thread in ductile material by pushing unwanted material aside, compressing rather than removing it. Pressure, friction and the resultant heat cause the metal to flow as the thread is formed, leaving a

> **GEOMETRIC (PROGRESSIVE) TAPS**
>
> A 'normal' tap (*see* Fig. 8-4) cuts a thread in one pass. There are usually three taps in a set of standard hand taps, but the difference is simply that although each cuts a full threadform, they may not fully cut some of the threads at the bottom of a hole. The first (or taper) tap leaves the bottom six less fully formed, the second tap leaves the bottom two or three not fully formed, and the third (or plug) tap leaves just the bottom thread not fully formed.
>
> In a set of 'geometric' or 'progressive' taps (*see* Fig. 8-4), the first tap cuts a partial threadform along the entire length of its cut, while the second tap finishes the whole threadform. This means the force exerted by each geometric tap is much lower than for a 'normal' tap. However, while the threads in a through hole could be fully formed by any single standard tap, fully forming a thread using the geometric taps always requires all the taps in the set to be used in sequence.

> **SPIRAL MACHINE TAPS**
>
> Spiral machine taps (*see* Fig. 8-4) have deep spiral flutes designed to eject chips efficiently. They are used for tapping using machines such as drilling machines, milling machines and lathes. Machine taps cut the whole of a threadform in one pass, and self-ejection of chips allows this kind of tap to be used effectively in a blind hole.

stronger and often tighter-fitting threaded portion. Fig 8-4 shows a small hybrid tap which has flutes only at the tip.

A self-tapping screw dispenses with the need for a tap to cut a thread. Instead, the screw drills its own hole and makes the thread as it is screwed into the work. Thread-cutting self-tappers often have a gash in the screw thread that acts like a tap, cutting a thread in the sides of the hole produced by the screw (Fig. 8-6). Thread-forming self-tapping screws act like a roll tap, forming the thread by pushing material aside rather than cutting. Self-tapping screws normally require a small pilot hole so that the tip of the screw is located in the correct position on the sheet; this should be no larger than the root diameter of the screw, and can be a little smaller.

FIXING TO SHEET

Captive Nuts

When the threaded portion in a sheet is not strong enough to allow a screw to exert sufficient grip, but there is insufficient access to allow a nut and bolt to be tightened, a captive nut can be fixed to a sheet by welding, soldering or gluing. Captive nuts can be held, loose, inside a cage fixed to a part. In a standard 19in equipment rack, for example, nuts are trapped inside a thin sheet cage which has springy ears on two sides (Fig. 8-7).

Fig. 8-6: Some self-tapping screws have a gash in the threads at the start, to help cut the threads as the screw enters the material.

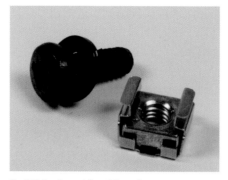

Fig. 8-7: Captive nut for a 19in rack.

To secure the nut to the rack, the ears are sprung into slots in the sides of the rack. To prevent sore fingers, a simple tool made from a bent strip of sheet metal is used to lever the ears into the slots. This tool is also used to remove caged nuts from the rack slots.

Once the caged nuts have been fitted, machine screws are then used to hold equipment in the rack. These screws pass through holes or slots in the front panel of the equipment housing and into the caged nuts. This holds the equipment securely, even in racks made of relatively thin sheet (Fig. 8-8).

Attaching a nut behind a sheet, whether by using a permanent fixing method or a cage, is a method commonly used where a screw will be repeatedly removed and re-inserted, to allow disassembly and reassembly of a set of parts.

In some situations where a hole will be near the edge of a sheet, cost can be reduced but grip maintained by using U clips: these are U-shaped plates in which one side has a clearance hole and the other side a 'thread' formed from a punched hole and two flanges (Fig. 8-9). The sheet should have a clearance hole. The clip slides over the side of the sheet and a component can then be secured to the sheet by using the clip as a captive nut.

One limitation is that the parts being held together cannot be closer together than the thickness of one side of the U clip. Note, too, that the clips tend to tilt the mating parts,

Fig. 8-8: Equipment secured using rack nuts and screws.

so that this must be taken into account at the design stage. Nevertheless, U clips are quick to use, and allow a useful degree of self-alignment so that some adjustment of the relative position of a panel and a component is possible during assembly. Where the location of a part is fixed by some other feature such as an edge or a pin or another component, the clip will align itself to the screw on assembly, so it allows more leeway than a soldered or welded captive nut. U clips are available in a variety of forms to suit particular applications.

Rectangular flat nuts can be located anywhere on a panel, and are not restricted to locations near the edge of a sheet (Fig. 8-10). These act as replacements for threaded nuts. However, securing a rectangular flat nut either requires access to both sides of the sheet, so that the nut can be positioned and the screw inserted, or the nut must be held in a captive arrangement of some sort.

U nuts are captive nuts that can be held in a cage resembling a U clip (Fig. 8-11), and slid in from an edge to lie under a hole. Like

Joining • 87

Fig. 8-9: U clips have a clearance hole in one side and a thread formed from a punched hole and two flanges in the other.

Fig. 8-11: U nuts hold captive nuts and, like U clips, can be located by pushing them over an edge.

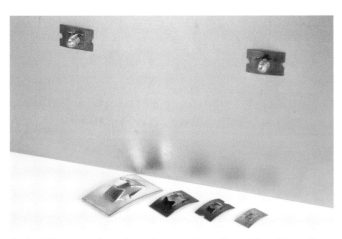

Fig. 8-10: Flat nuts can be used away from an edge and have a 'thread' formed from a punched hole and two flanges.

Fig. 8-12: U nuts can also be used in the centre of a panel by punching a slot and creating a bent tab.

the U clip, these can easily be used near the edge of a panel. However, they can also be inserted away from an edge by creating a small rectangular hole with an angled entry and a bent tab at the rear (Fig. 8-12). Once inserted, the bent tab can be flattened to lock the clip in position, if necessary. The U clip can also be used this way.

Clips

Clips allow assembly and, in some cases, disassembly of one sheet component to another, of decorative materials (such as car trim panels) to sheets, and of cables to sheets, and can secure rods and tubes in position. Clips are available in a wide range of sizes and styles to suit a variety of applications (Fig. 8-13), and can hold sheet flanges

Fig. 8-13: Attachment clips are available in a wide range of styles and sizes, and can hold panels together at various angles.

Fig. 8-14: Where a clip is designed to attach a decorative panel to a sheet, the side against the decorative finish will have no barbs, to avoid damaging the panel.

together, or one sheet aligned to another, or one sheet at right angles to another, depending on the orientation of the slots in the clips.

Normally, clips incorporate barbs that bite into the sheet and help hold the clip and the sheet parts in position. Where a clip is designed to hold a part made of soft material, such as cloth or leatherette, or a part with a decorative finish, at least one part of the clip will have no barbs, to avoid marking the finish (Fig. 8-14). This often allows careful disassembly, especially of car trim panels where a bodywork repair is required.

Cable edge clips can also hold cables in position on a sheet, and are available in forms that can hold single or multiple cables (*see* Fig. 8-13, top row).

Rivets

Rivets provide a strong, permanent way of assembling parts (Fig. 8-15), as well as being decorative. Although there are many types of rivet, the basic form consists of a shaft (or shank) with a head at one end. The rivet is inserted into a hole through two or more sheet components, then the end (or 'tail') of the rivet is deformed by spreading it and causing it to squeeze the components together (Fig. 8-16). The length of the rivet is related to the thickness of the parts being held together, and also to its diameter, so that when the tail is deformed it is able to spread sufficiently to provide a good grip on the components.

The head of a rivet may be designed to protrude above the finished work, or may be countersunk or flat (Fig. 8-17).

The tail of the rivet may be accessible on the other side of the sheet assembly, or may not, in which case the rivet must be designed for one-sided assembly. The top sheet component may be assembled against a much thicker part, and the rivet may not be able to reach right through, so it must be designed to work in a blind hole.

Fig. 8-15: Rivets serve a functional purpose in joining parts together permanently, but they can also be aesthetically pleasing. Photo of the Carl Webb Gilera 500 replica.
BRIAN WALBEY

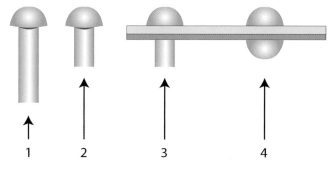

Fig. 8-16: Stages in riveting.

1 Select a style of rivet

2 Trim the shank to suit the job

3 Insert in a hole through the parts to be joined

4 Clench the rivet by forming a head on the reverse side

Joining • 89

Fig. 8-17: Rivets are available in a range of materials, with different head styles, shank diameters and lengths.

The diameter of the shaft of a rivet should be chosen to cope with the forces acting on the rivet in the joint.

The material from which the rivet is made may need to be chosen to suit the components being held together, so that, for example, the rivet does not corrode or cause corrosion. Copper riveted to cast iron, brass to aluminium, brass to galvanized steel, and plain low-carbon steel in contact with stainless steel are all likely to result in corrosion, as are many other combinations of metals.

All these factors combine to influence the design choices and to determine the size of the rivets used in a joint.

Application of Rivets

When assessing the dimensions of rivets for a particular job, the diameter of the rivet shaft should be at least three times the thickness of the thickest sheet, or as specified on engineering drawings. The hole for the rivet shaft should be $D \times 1.0625$ (or $D \times 1\frac{1}{16}$), where D is the diameter of the rivet shaft. Alternatively, make the hole 0.075mm larger than the rivet shaft. When pushed through the hole, the rivet shaft should project 1.5 times the diameter of the shaft, although experts may be able to set the rivet with a projecting length as little as $1.25 \times D$.

This will allow the tail of the rivet to be formed into a shape with a width at least 1.5 × the rivet shaft diameter, and a thickness at least half the diameter of the rivet shaft.

However, if there is too much or too little

Fig. 8-18: Combined setts (the plain hole) and snaps (the domed recess) for each of three sizes of rivet.

material projecting, it will be difficult to form a neat head on the rivet.

Clenching the rivet by hammering the tail will squash it and spread it at the same time, squeezing the components together and preventing them from being pulled apart. As the tail is struck, the same force acts on the head at the other end of the rivet, deforming it, too. The original shape of that head is important, because its proportions ensure appropriate strength and resistance against pulling through the sheet. So it is usual to support the head in a shaped recess.

Tools to shape the tail (Fig. 8-18) can give a consistent pleasing finish. This involves striking the tail with a shaped punch, usually to form it into a domed shape. For all but the smallest rivets, additional punches help in the intermediate stages of forming the heads (Fig. 8-19).

To put in a rivet, proceed as follows:

◆ Support the pre-formed rivet head in a shaped recess in a plate, a dolly, or a stake. The dolly should rest on a solid object such as a large anvil or block of steel, to absorb the hammer blows, or be held in a staking tool (Fig. 8-20). The tail end of the rivet shaft will now be pointing upwards
◆ Place a set tool over the shaft, and strike it firmly to make sure the rivet and the material are sitting in contact and in the right position. The set tool simply has a hole which is an easy fit for the rivet shaft. Tapping the tool seats the components together
◆ Strike the end of the rivet shaft firmly with a hammer or a cone tool to begin forming the shape of the new head on

Fig. 8-19: Left to right: a rivet set (sometimes spelt 'sett' for historical reasons), a flat tool (flat-ended punch), a cone tool (conical punch), a snap (rivet head punch), and a stake to support a rivet head in a staking tool.

Fig. 8-20: A rivet balanced in a staking tool, for illustration, with a stake underneath to support the head, and a punch above, to form the tail.

Fig. 8-21: The tail of a rivet after the cone tool has been used.

final shape. The snap should have an internal shape the same as the final shape to be formed. Alternate the use of the snap tool with blows from the hammer or from a flat tool (a flat-ended punch) applied to the top of the head being formed. Finish by using the snap tool

Use of the cone tool is often omitted, but forming a cone first does help the process of forming the final shape, especially on larger rivets, and ensures that the new head on the rivet will sit down tightly on to the components being riveted together.

Use of the flat tool is also often omitted, but its purpose is to produce a flat at the top of the head that will prevent energy from the impact of the snap tool being dissipated by contact with a large surface area. Keeping the top of the new head clear of the snap will allow more energy to be put into forming the sides of the new head, and allow metal to flow into the final shape more easily. At the end, a few additional blows on the snap tool will finish the shape to the full size (Fig. 8-22).

If the rear of the rivet will not be seen, the finished shape of the tail will not be important, so the snap may be omitted and the

Fig. 8-22: Finished rivet tail.

- the end of the rivet shaft (Fig. 8-21). The recess in the cone tool is a truncated cone (that is, a cone with a flat at the smaller end)
- ◆ Position a snap over the partly formed end of the shaft, and hit it with a hammer to begin forming the

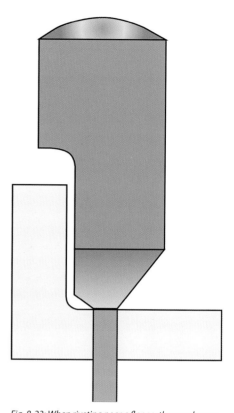

Fig. 8-23: When riveting near a flange, the punch may need to be relieved on one side.

final shape formed using the hammer or a flat punch, as long as the finished shape will be large enough and thick enough to prevent the rivet from pulling through the components.

Straight cylindrical punches can be used, except where a rivet is close to a flange (Fig. 8-23), when the punch may need to be relieved on one side.

The dolly can be a depression in a plate, or in the end of a cylindrical rod, but it may need to be offset at the end of a bent bar where access is restricted (Fig. 8-24). Offset dollies quickly become too flexible to be effective.

Some rivets may be set using a pneumatic squeezer (Fig. 8-25) designed to close double-countersunk rivets, as used on some aircraft. Aluminium rivets already have one countersunk head, but the flat faces on the ends of the arms of the closer exert a sufficiently high pressure to form the second

Joining • 91

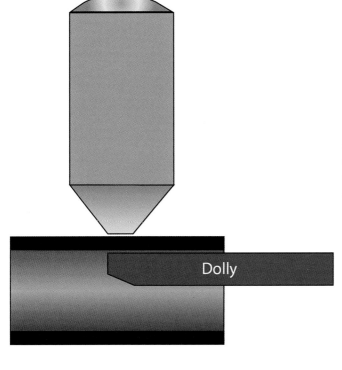

Fig. 8-24: An extended dolly may be needed to reach inside work or beyond an obstruction.

Fig. 8-25: A pneumatic tool for clenching flat-headed rivets by giving them a flat tail.

countersunk head. Using a mechanical closer ensures consistency over a large number of rivets.

Blind Rivets

Where the reverse side of the work is inaccessible, and a rivet cannot be closed because the set and snap punches cannot reach the rivet, a blind rivet must be used.

Blind rivets take several different forms, but are all inserted then finished to shape from one side of the work. The most common and most versatile version of the blind rivet is the 'pop' rivet (Fig. 8-26). In its modern form,

a pop rivet is a hollow tubular rivet, often made of aluminium, which has a mandrel (a short straight shaft) running through the rivet body. The rivet and mandrel are

Fig. 8-26: Pop rivets and rivet pliers.

Fig. 8-27: The corner of a tray, secured by a pop rivet.

inserted through a hole in the work, then the mandrel is pulled to squeeze the rear of the rivet against the work. When the force reaches a pre-determined amount, the mandrel snaps, leaving the head trapped in the rivet (Fig. 8-27).

Pop rivets are usually set using a mechanical device to pull the mandrel, and this normally takes the form of a collet that closes on the mandrel shaft. The puller may be squeezed like a pair of pliers (Fig. 8-28), or pumped towards the work (Fig. 8-29), or it may be operated by air pressure (Fig. 8-30).

Variations on the standard pop rivet includes threaded inserts, termed Rivnuts. These are inserted and secured in place using a threaded mandrel which does not break but is unscrewed once the rivet has been clenched, leaving a threaded hole for a screw or bolt (Fig. 8-31).

Because pop rivets are set using a simple mechanical action, they are easy to use. They are available in aluminium, steel, stainless

Fig. 8-28: Long-handled rivet pliers used to clench the larger pop rivets.

Fig. 8-32: A pop rivet can be set with a large washer under its head to spread the pressure when gripping soft materials.

Fig. 8-29: Lazy tongs: concertina-style pop-rivet pliers that work using a pushing rather than a squeezing action.

For this technique to be effective when riveting two soft materials together, the rear side of the work must also be accessible. A washer is then added under the rivet head, the rivet is passed through the work, and a washer is slipped over the end of the rivet, before setting the rivet and trapping both washers next to the work (Fig. 8-32).

Despite the need to set the tail end of the pop rivet, specially designed forms of pop rivet can be used effectively in blind holes. The grooved pop rivet, for example, is an aluminium rivet designed to grip inside blind holes in wood or similar soft materials. When the rivet is set, the grooves allow the rivet to collapse while expanding sideways against the sides of the hole.

Other specially designed pop rivets include rivets that seal when set, and are water- and gas-tight.

material, pop rivets are available with large-diameter heads to spread the load under the head and prevent the material being crushed, or the rivet pulling through the soft material. Where these are not available, one popular strategy for spreading the load under the pop rivet head is to arrange for the soft material to be under the rivet head on the accessible side of the work and to add a large-diameter washer under the rivet head.

Fig. 8-31: Rivnuts and a closing mandrel which can be gripped and pulled by pop rivet pliers.

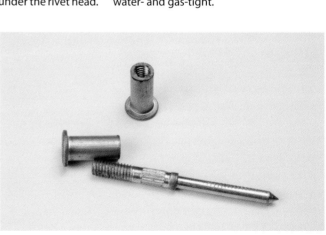

Fig. 8-30: A pneumatic pop rivet closing tool.

steel, copper and monel metal (an alloy of nickel and copper, noted for its resistance to acid corrosion).

When joining a soft material to a harder

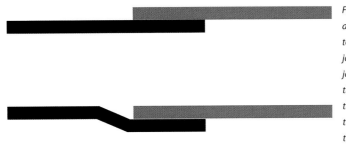

Fig. 8-33: The effect of a joggle is to bring the top surfaces of two joined sheets level. The joint remains two sheets thick, but the change in thickness is on one side of the joint and only affects the joint area.

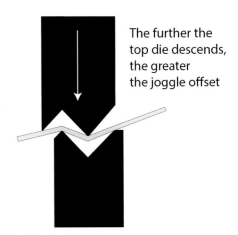

Fig. 8-34: A joggle can be produced in a press brake using appropriate tooling.

Joggling

When one sheet is joined to another, this often results in two thicknesses of sheet at the join (Fig. 8-33). The consequence is that one sheet is then not aligned with the other, but is at least one thickness out of alignment. This can be avoided by forming an offset step or joggle in one of the sheets (see Fig. 8-33). Joggles are commonly just one sheet-thickness deep, but they may be up to five thicknesses of sheet in depth; beyond that they are simply offset bends. Joggles may be formed in a jenny or bead roller (see Fig. 7-14).

Because the depth of a joggle is relatively small, it is difficult to form in a folder, but it is possible in a press brake if the die has a thin lip (Fig. 8-34). One easy way to form a joggle at the edge of a sheet is to use a joggle tool, available either as a manually operated tool (Fig. 8-35) or, for longer sections, a pneumatic tool (Fig. 8-36). These are essentially press tools, with an upper tool and a lower die forming an L-shaped step. Place the sheet between the jaws and squeeze to form a single-thickness joggle consisting of two bends.

To join two sheets, joggle one, then place the other in the step; then secure with screws, rivets or spot welds. The face with least evidence of the joggle is called the 'show' face, and the joggle flange lies behind that face, giving a neat appearance. This is a technique often used to join sheet metal parts in automobiles, as it lends itself well to spot welding.

Fig. 8-35: Joggler/flanger and punch tool. JACK SEALEY LTD

Fig. 8-36: Air punch/flange tool. JACK SEALEY LTD

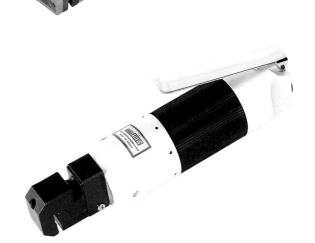

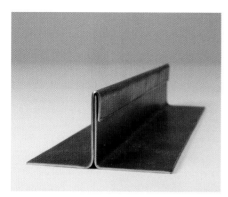

Fig. 8-37: A folded edge holding a second sheet part in position.

FOLDING TOGETHER

A simple fold can be used to hold two sheet parts in place (Fig. 8-37). Two flat sheets can be joined using two folds, one at the edge of each sheet (Fig. 8-38). Fold the ends of two sheets into thin 'U' sections, place one end inside the other, then close them by hammering or pressing. The narrower the 'U' the better the end result, and the less each end 'grows' as the bends are collapsed. Depending on the design of the parts, and the way they are fixed to other components, it may be enough to fold the ends together like this. However, if the parts need to be held together, the clenched U folds may need to be pinned, screwed or riveted together through the four-folded thicknesses.

To make a simple U fold, use a folder to make a bend greater than 90 degrees, then close that fold, in the same way as for a hem (a folded edge), but close it loosely over a single thickness of the mating sheet. Do the same for the other sheet. The joint can then be assembled and closed. The result is two joined sheets offset by the thickness of the folds. The joint itself will always have a thickness of at least three sheets, but the offset between the main sheets can be reduced as required by incorporating a joggle (Fig. 8-39).

Making a more complex joint based on the same principles can help lock the sheets together, preventing movement in one or more directions. Fig. 8-40 shows a number of different possible joins whose final assembly involves folding, pressing or hammering the joint closed.

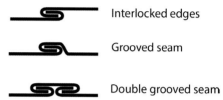

Common interlocked joints

Fig. 8-40: Commonly used interlocked joints made by folding and hammering. Security of the joint varies from low (interlocked edges) to high (double-grooved seam) as possible movement of the sheets is increasingly restricted.

Fig. 8-38: Two sheets can be joined by interleaving U bends at adjacent edges.

Fig. 8-39: Adding a joggle alters the offset between the surfaces of the sheets. The joint shown offsets the surfaces on either side of the joint by one sheet thickness, but this can be adjusted by altering the joggle.

Several of these joints are difficult to make using a folder or a press brake, and the lock-formed joint is best made using a dedicated machine, as are some of the others. A lock former is a roll-forming machine consisting of a set of forming rollers, like the rollers used in a jenny, which gradually form the edge of a sheet into the required shape to mate with another. Sheets can then be assembled and closed.

Different profiles can be used to join sheets side by side, or at right angles.

SOLDERING

Solder is a filler metal that melts at a lower temperature than the metal to be joined. Heating the joint so that the solder melts and can flow into the joint, then allowing the joint to cool, results in a permanent mechanical joint being formed which has some strength and may be able to seal the joint against fluid leakage (Fig. 8-41).

Fig. 8-41: Soldered joints.

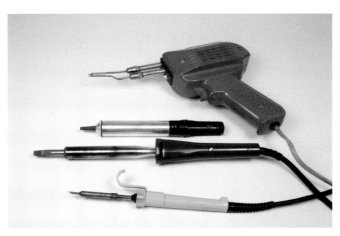

Fig. 8-42: Soldering irons with different heat outputs, and a solder sucker tool.

Soldered joints usually have good electrical conductivity.

Soldering in all its forms involves heat and acidic fluxes as well as acid washes, so all forms of soldering require appropriate safety measures, especially protection against chemicals and hot soldering irons or gas torches, as well as adequate ventilation to prevent the inhalation of vapours which may contain traces of metal, acid fumes, powdered or liquid acidic fluxes.

Soft solders are traditionally an alloy of tin and lead, but modern lead-free soft solders are alloys of tin, antimony, bismuth and silver and have melting points between 138 and 300°C. Soft-soldered joints can be made using soldering irons (or 'bolts') normally heated by electricity or gas (Fig. 8-42). Absolute cleanliness of the component parts is required for a successful joint, so in order to remove oxidation, dirt or grease from the area to be joined, a flux (Fig. 8-43) is used to chemically clean the metal around the joint, and the flux is left on the material while the joint is being made. The metal is heated by a soldering iron, then solder is applied to the joint. The solder melts, spreading into the joint by capillary action, then the soldering iron is withdrawn. The joint cools and the solder solidifies, creating the mechanical joint.

A variation in this method is to coat each part with solder before the joint is made, by heating and allowing solder to flow over the area to be joined. This process is called 'tinning', probably because solder is an alloy of tin. Assemble the parts together with the tinned surfaces in contact, then heat the joint area until the solder re-melts and forms the joint.

Silver solders, or 'hard solders', are alloys containing a high proportion of silver (typically 56 to 80 per cent) with a melting point in the range 618–740°C (Fig. 8-44). Because it requires a higher temperature and a greater amount of heat to make a silver-soldered joint, heating is normally by

Fig. 8-43: Soft solder in wire and stick form, and tins holding different grades of soldering paste.

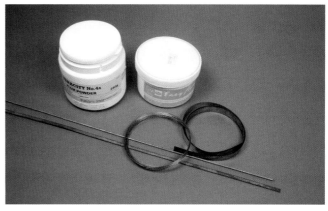

Fig. 8-44: Silver (hard) solder in strip, rod, wire and foil form, and silver-soldering pastes.

Fig. 8-45: Oxyacetylene torch, copper-coated steel filler rods, brass filler rod and flux used for brazing and bronze welding as well as silver (hard) soldering.

gas torch (Fig. 8-45). Burning gas tends to accelerate the oxidation process as well as introducing more impurities into the area to be joined, so using a flux is essential. Most fluxes for silver soldering are supplied in the form of powder, to be mixed to a thick paste by adding water.

Clean the joint area and apply flux. Heat the area as quickly as possible until the flux flows and the joint reaches the melting point of the solder. Then apply the solder: it should melt and flow into the joint by capillary action. Remove the heat and let the joint cool. Once cool, remove the flux residue. This may take the form of a glass-like film, and the whole job may need to be soaked in water, acetic acid (vinegar) or an aggressive acid such as battery acid or mild sulphuric acid. Take care when handling acid: follow the instructions given by the manufacturer or supplier, and wear appropriate protective clothing and a mask as well as ensuring adequate ventilation. Wash the job thoroughly afterwards.

A silver-soldered joint is mechanically stronger than the equivalent soft-soldered joint, but requires more heat and uses a more expensive filler metal.

Brazing is like silver soldering, but the filler material is usually a brass alloy rather than a silver-based alloy. Otherwise the process is similar.

Soldering, silver soldering and brazing use high temperatures to allow the filler alloy to melt and flow into the joint, but none of these processes involve melting the parent material (the workpieces being joined).

WELDING

Welding involves heating two (or more) metal workpieces until the joint area melts and the metals fuse. A metal filler is often added at this stage, by melting it into the joint.

There are several different methods of melting the metal in the joint area.

A gas torch (*see* Fig. 8-45), which normally burns a mixture of oxygen and acetylene, can be used to melt the work in the joint area to form a molten pool. A filler material can also be melted into the weld pool to strengthen the joint. Removing the heat allows the weld pool to cool, and the resulting joint will have a mechanical strength similar to the parent materials. This form of welding is often known as oxyacetylene welding. It is a popular form of welding but requires some skill in forming and manipulating the weld pool without allowing the pool to collapse, forming holes in the work. The heat involved in this process is considerable, and that tends to lead to distortion, especially in thin sheet. On an area such as a thin car-door panel, for example, it is very difficult to avoid distortion, despite the use of cold wet cloths or heat-conducting paste.

Aside from welding, oxyacetylene gas torches are a useful source of intense heat for annealing, bending and forming metals.

Electric Arc Welding

In electric arc welding, also known as stick welding or manual metal arc welding (MMA), an electrical current is passed through a rod of flux-coated filler metal, into the joint (Fig. 8-46). An arc forms between the end of the filler rod and the joint metal, and the high temperature melts the end of the rod and the joint material as well as the flux coating on the rod. Droplets pass from the end of the filler rod into the molten weld pool, while the molten flux prevents contamination of the weld area.

This is a well-established method of welding, based on simple equipment, but although physical control of the rod and electrical control of the current allow the method to be used on a wide range of thicknesses of parent material, control

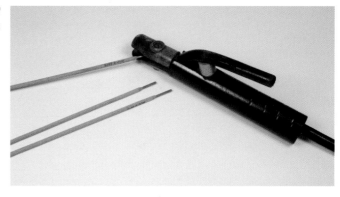

Fig. 8-46: MMA (stick) welding rod holder.

becomes much more difficult as the materials become thinner; it is therefore not a particularly good method to use on thin sheet. Its main applications lie with materials more than 2 or 3mm thick.

Metal/Inert Gas (MIG) Welding

MIG – metal/inert gas – welding is a method of electric arc welding in which the flux is supplied as a gas, instead of as a covering on the welding rod. The weld filler material is supplied as a wire, which is fed into the joint area from a reel in the welder, so that the operator does not have to stop the weld to change the filler rod. A welding 'torch' or 'gun' is used which takes the form of a handle with a shroud, an inner nozzle and a trigger (Fig. 8-47). Pulling the trigger causes gas to flow through the shroud, surrounding the joint area in front of the torch. Wire is fed through the nozzle towards the joint.

When the wire touches the joint metal, a current flows, which heats the weld area and melts the tip of the wire. As the wire melts, it loses contact with the weld pool and an arc forms, as in stick welding. Wire continues to be fed towards the weld pool by a motor in the welder, and the rate of feed matches the rate at which the wire is melting; this maintains the arc, and welding can continue as long as wire is being fed through the torch.

MIG welding is easier to master than stick welding, particularly when striking the arc initially, but both use the same principle. Variations on the basic equipment required include gasless MIG welders which feed a flux-cored wire to the joint, dispensing with the need for a separate supply of gas.

MIG welding causes less heat to be created across the joint area than gas welding, and is more suited to welding thin sheet. Because control of the arc, the wire and the gas is carried out by control electronics in the welding unit, this method is more reliable and consistent than stick welding, and is often easier to use.

Tungsten / Inert Gas (TIG) Welding

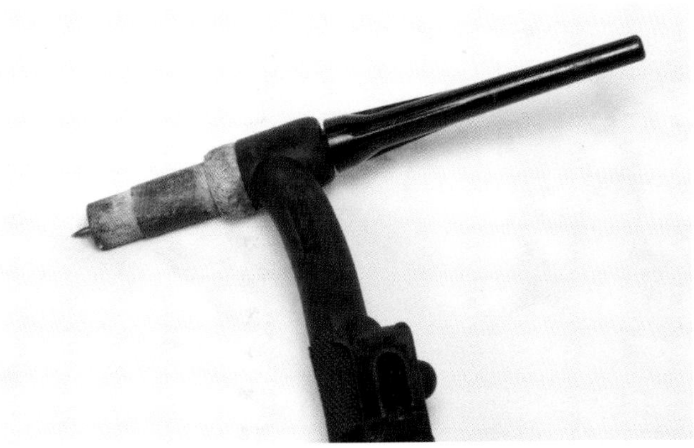

TIG (tungsten/inert gas) welding combines the techniques used in gas and MIG welding. Instead of feeding a filler wire through the welding gun, the inner nozzle is replaced by a fixed tungsten electrode (Fig. 8-48). Filler wire is introduced to the joint manually, in the same way as when using gas welding. Pulling the trigger on the gun activates the welding control unit and causes shielding gas to flow through the shroud and envelope the weld area, but the initial arc must be struck by touching the electrode to the work, in the same way as for stick welding. Once the arc has been struck, and a weld pool formed, filler rod is introduced manually as and when required.

TIG welding requires more skill in forming the arc and manipulating the weld pool. It is also slower than MIG welding, but it does allow the kind of delicate control of the weld pool normally associated with oxyacetylene welding, so in some ways it combines the best characteristics of several techniques.

Spot Welding

Spot welding is a form of electric resistance welding, in which an electrical current is passed through two metal surfaces in contact with one another. The electrical resistance across the joint creates heat, melting the surfaces at the point of contact, and so welds them together. A spot welder (Fig. 8-49) uses two copper electrodes to press the workpieces together, and the combination of heat and pressure produces a weld that resembles a large spot (Fig. 8-50). The process only takes a fraction of a second, so spot welding is a quick way to weld sheets together. The current, the time for which the current will be applied, and the pressure applied by the arms are

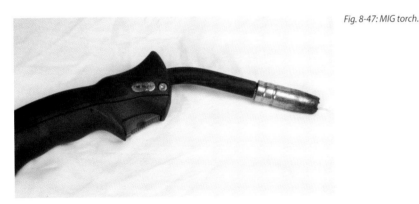

Fig. 8-47: MIG torch.

Fig 8-48: TIG torch.

98 • Joining

Fig. 8-49: Spot welder.

Fig. 8-50: Spot welds. The central silver spots are where pressure and current are applied to make the welds. The blue circles are heat marks.

Fig. 8-53: Fitted with appropriate tips and arms, a spot welder can reach into a narrow gap.

Fig. 8-51: Pressure adjustment under the top handle.

Fig. 8-54: Selection of spot-welder arms and tips.

Fig. 8-52: Timer adjustment.

Fig. 8-55: Examples of typical spot-welded joints.

electrodes on either side of the joint, and squeeze the handle of the welder (Fig. 8-53).

Spot welding is an ideal way to join one sheet component to another, and is widely used in sheet metal work. Spot-welding machines are typically used with various lengths of arm, which can carry different shapes of electrodes, so that the welder can reach the spot to be welded despite obstructions. Fig. 8-54 shows a typical selection of arms, and Fig. 8-55 shows a selection of spot-welded joints.

GLUED JOINTS

Modern adhesives are strong enough to be used in many applications where welding would formerly have been required. These adhesives work at normal room temperature, so there is no risk of distortion because of the heat of the welding process. Adhesives do need to be chosen with care, however, so that they are appropriate for the particular joint to be made

all adjustable (Figs 8-51 and 8-52), allowing fine control over the quality of the weld.

To make a spot weld, adjust the current, time and pressure, then position the

– but that is a little like choosing the best solder or welding process for a joint. Panel adhesives can be used to eliminate traditional fixings such as screws and rivets, as well as welds. They allow dissimilar metals to be joined easily, and they allow dissimilar materials to be bonded, such as plastic to metal.

Typically, these adhesives are either acrylic compounds that are mixed together just before pressing two panels together, or resin adhesives that are used in much the same way. Once the parts have been clamped together, the adhesives must be left for some time (perhaps an hour or so) to set, and achieve full strength: this will take twelve to twenty-four hours, depending on the type of adhesive. On the other hand, bolting, riveting and welding achieve full strength immediately, or as soon as the work cools after welding. Nevertheless, panel adhesives create a strong bond, and are a good way of securing one sheet metal panel to another.

Consult the manufacturer's literature for the exact specifications and for guidance as to the best adhesive to use on a particular job.

CLAMPING DEVICES

Clamps are useful for holding parts together while a joint is made. Although many commonly used clamps and grips can be used on sheet metal (Fig. 8-56), some self-gripping pliers specifically designed for sheet metal work are available (Fig. 8-57).

Cleco fasteners (Fig. 8-58) or skinpins are spring-loaded pins that are split at the lower end and divided by a tongue. These pins are designed to be inserted through a specific size of round hole in sheet components, and to grip from behind, holding the components in position until a sound joint can be made. The pins are then removed and the holes sealed.

Loading a pin into the purpose-designed pliers allows the spring to be compressed and the split end to be pushed forwards (Fig. 8-59). The shape of the ends allows them to be pushed through a hole, and releasing the pliers causes the split ends of the pins to be retracted but spread on either side of a central tongue, forcing them outwards to grip the sheets (Fig. 8-60).

Fig. 8-58: Cleco fasteners and operating pliers.

Fig. 8-59: Cleco being operated by pliers, showing the bottom barbs being pushed out beyond the central tongue to allow them to move closer together so that the fastener can slip into a hole.

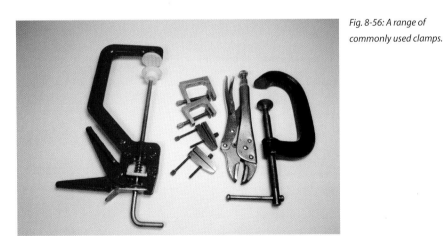

Fig. 8-56: A range of commonly used clamps.

Fig. 8-57: Self-gripping pliers designed for sheet use.

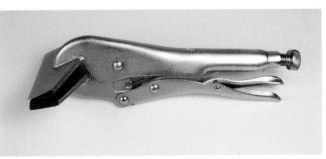

Fig. 8-60: Clecos in use holding two parts together.

9 Forming, Pressing and Drawing

Changing the shape of a flat sheet by cutting, folding, rolling and making holes in it allows the production of a wide range of sheet metal parts, but there are times when the sheet needs to be stretched, shrunk or formed into shape in a controlled manner. We won't deal with the more artistic free-form beating of metal into 3D curves in this section, but will concentrate on the more predictable methods of forming, such as doming and drawing – though some say that's just as artistic.

USING FORMERS

A plain sheet can be persuaded into a more complex shape by forcing it over a former. The former may be something as simple as a circle of steel (Fig. 9-1) with a radiused circumference, and the sheet may be forced over the former using a mallet. For example, a simple flat circle of copper can be made into a flanged end for a boiler (Fig. 9-2) by clamping the sheet against the former (Fig. 9-3) and beating the edge down over the former, in stages. Copper is a soft metal but it hardens as it is worked, so it may be necessary to anneal the sheet partway through the job, and return it to a soft state.

To anneal copper, heat it to a dull red and leave it to cool. Complex formers with tight curves may need the copper sheet to be annealed more than once during the course of a job.

The sheet is held against the former by a backing piece as the flange is being formed, and that backing may be another shaped steel former, or a piece of thick wood or MDF. For a simple shape a backing is not strictly necessary, as it is easy enough to flatten the sheet against the former, using the mallet. However, a backing does allow the sandwich of former, sheet and backing to be clamped in a vice to provide support as the edge is hammered over the former.

Hardwood formers work well for thin, soft sheet, but a steel former is more durable and more able to withstand the stresses involved in shaping a thick sheet.

To shape a sheet, proceed as follows:

- Begin with former and blank clamped together, or held between plain vice jaws. Using either a hardwood mallet with a flat face, or a flat-faced steel

Fig. 9-2: A copper boiler end plate shaped from a flat blank by using a steel former plate.

Fig. 9-1: A steel former plate.

Fig. 9-3: Clamp the former plate and the blank.

OPPOSITE PAGE:
A combined blanking and deep drawing tool.
JOHN SAUNDERS AT NYC CNC: WWW.NYCCNC.COM AND ON FACEBOOK

Fig. 9-4: First stage in forming the blank.

Fig. 9-5: After annealing the blank, tap it into close contact with the plate, then use a planishing hammer to finish the work.

hammer, strike the edge of the plate at a flat angle, bringing it towards you and downwards with a slight 'wiping' action, to draw the edge of the blank forwards and down on to the former. It will not seat fully and finally at this stage, but by striking round the circumference perhaps twice, it will reach the stage shown in Fig. 9-4.
- Work in sections, rather like stitch welding, and do not try to work continuously around the circumference (*see* the comments on shrinking, below). At each section, strike the first few blows on one spot, then work outwards, first to one side, then the other. Finally, work around merging any discrepancies between one section and another.
- Some say this work is too heavy for a mallet, preferring a steel hammer, and it is true that repeated blows right on the sharp edge of the blank will tend to damage the mallet – although it does have the advantage of a softer face. It's a matter of personal choice.
- Anneal the blank and leave it to cool. Alternatively, anneal the blank and quench it in water: that cools it more quickly, and tends to loosen the scale formed during annealing.
- Using a planishing hammer with a polished flat face, tap the edge down tight on to the former with firm, but not heavy, strokes. Some blows around the flat face of the sheet, just at the top where it begins to turn over the former, will help ensure it is tight and flat at the front as well.
- Then planish the edge using rapid and very light blows delivered by a flick of the wrist, with the shoulder and biceps relaxed. Planishing smooths, heavy blows stretch as the blank is trapped between hammer face and steel former. The end result should be as shown in Fig. 9-5.
- Apart from a few blows near the turn from the face on to the side, over the curve on the former, the circular front face should not be touched by hammer blows and should remain unmarked.

Shrinking

The circumference of the outside of the flat disc is greater than the circumference of the former, so that edge must be shrunk as it is turned over the former. With a soft blank, shrinking will take place almost automatically. Working in sections avoids any excess material being pushed around the circumference, creating ripples.

If ripples become a problem, the edge can be shrunk as required. It is too awkward to do that using a stretcher/shrinker, so a shrinking hammer can be used instead. The shrinking hammer has a chequered pattern on its face, and that acts to drive some points on the sheet downwards and inwards, while allowing the adjacent points to expand into the gaps. The effect is to shrink the metal slightly. Any resulting marks can be eliminated by planishing, at the last stage (*see* below).

Planishing

Once the edge has been formed, its surface can be tidied up using a planishing hammer, which has a smooth face. Planishing uses many light blows, because heavy blows cause stretching as the metal is squeezed between the hammer and the former. Using many light blows tends to smooth the surface and provide a good finish.

For larger workpieces, planishing can be automated by using a powered planishing hammer, or a converted air chisel to deliver rapid light blows to the surface.

Although this is a hand process, the former provides sufficient guidance that the end result is easily repeatable and accuracy is good.

PRESSING

Some shapes can be produced by pressing a sheet between a die and a former. A domed end plate, for example, can be produced from a flat circular blank by using a domed steel, plastic or hardwood punch to force the sheet into a matching recessed die. This can be done in a fly press or a hydraulic press, though some shapes can simply be formed by using the punch as a handheld tool (Fig. 9-6).

Fig. 9-7: Exhaust box flanges made from flat blanks by punching and flanging.

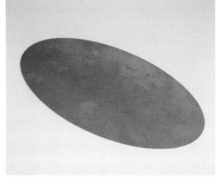

Fig. 9-8: Laser-cut blank.

Fig. 9-6: A doming block, doming tool with spherical head, and an aluminium item made from a flat blank by placing it in a recess and using a hammer to tap it down with a doming tool.

Dies can be made with locating pins or a recess to hold the blank securely and repeatedly, and the punch can be aligned with reference to the die, perhaps using spacers or feeler gauges, or by providing an auxiliary locating/guiding pin or other feature. It all depends on the accuracy required of the finished product.

Industrially, pressings can be relatively complex, and may be carried out in several stages, forming part of a shape, or specific features, at each stage. For more complex shapes, the grade of metal becomes important, as the material is subject to considerable repeated stresses. Punch and die making is a precise science, and the finished tool must take account of the thickness of the metal that will be trapped between the two, as well as necessary working clearances.

Exhaust box flanges as produced by RMS Engineering, for example, consist of a flat elliptical or circular plate with upturned edges, and a circular hole with upturned edges into which fits a matching tube (Fig. 9-7). The upturned edges simplify welding the flange to the pipe, or the flange to the outer body of the exhaust box.

♦ Begin with a flat elliptical plate, produced by blanking or laser cutting (Fig. 9-8). A preliminary operation is to punch a hole a little smaller than the exhaust pipe.

♦ The plate is then placed between locating pins in a flanging tool, and the central elliptical shoe is pressed down through the opening in the die, dragging the plate with it. The plate edges are upturned as it passes through the die (Fig. 9-9).

♦ Next, the hole is placed over a circular die (Fig. 9-10) and a punch is pressed through (Fig. 9-11), dragging the edges downwards, resulting in a flange around the hole. In this operation, the punch has a pilot diameter which locates in the plain hole to ensure an even flange.

Fig. 9-9: An elliptical flanging tool that will raise a flange around the outside edge of an elliptical blank.

Fig. 9-10: Punch and die for flanging the inside edge of a hole.

Fig. 9-12: Component parts of the NYC CNC blanking and deep drawing tool, showing the main body, blanking and drawing punch, a punched strip, flat circular blank, and drawn cup. JOHN SANDERS AT NYC CNC

Fig. 9-11: The plate is located by its outside flange, and the die is forced through the hole to produce a flange.

DRAWING

Drawing is rather like pressing, but the end result is deeper. In drawing, a flat sheet is changed into a relatively deep shape, in one or more stages. In most cases, as the shape is changed, the metal may also become thinner in places, as a result of the considerable stretching which takes place. A simple circle, for example, may be pressed into the shape of a deep cup, by pressing a cylindrical punch on the centre of the circle and dragging it down into a cylindrical die. Significant forces are involved, and the sides are shaped by the walls of the die, and the circumference of the punch.

Some sheet metals are available in a deep drawing quality, whose composition is specially designed to allow the considerable stretching and the high forces involved in this kind of process.

COMBINED OPERATIONS

Often, more than one operation can be combined to form a part. For example, John Saunders at NYC CNC has developed a simple tool that combines two processes:

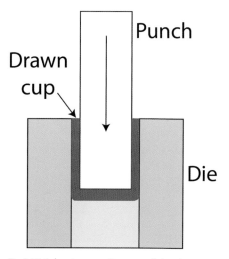

Fig. 9-13: A drawing operation normally involves pressing a punch against the centre of the blank, forcing it into a tube smaller in internal diameter than the blank.

blanking and deep drawing (Fig. 9-12). The first operation involves producing a circular blank from a strip of copper, using a circular punch and die. This uses the outer edge of the punch shown on the right in Fig. 9-12. Producing a blank involves the die being dead to size, matching the required diameter of blank exactly. The punch is slightly smaller, to provide a working clearance.

The subsequent drawing operation normally involves pressing a punch against the centre of the blank, forcing it into a tube smaller in internal diameter than the blank (Fig. 9-13). The excess material on the blank is forced down the inside of the tube, stretching but being compressed between tube and punch to form a cup shape.

In the NYC CNC approach, the two operations are combined in one tool, using a pin inside the lower part of the blanking die which acts as the punch, and the inside of the blanking punch as the die for the cup (Fig. 9-14). Fig. 9-15 shows the copper strip located in a recess to register it accurately for blanking (visible as the slot in the front of the tool in Fig. 9-12), and Fig. 9-16 shows the finished cup upside down on the lower punch.

Forming, Pressing and Drawing • 105

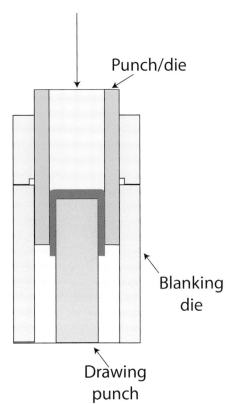

Fig. 9-14: The NYC CNC approach involves using a pin inside the lower part of the blanking die which acts as the punch, and the inside of the blanking punch as the die for the cup.

Fig. 9-16: The finished cup, on the top of the pin inside the tool. The copper strip is guided inside the upper part of the tool, for blanking. JOHN SANDERS AT NYC CNC

In the combined tool, the punch descends and produces the blank, then continues to descend, acting now as the die, and as the lower punch draws the blank into a cup shape.

The tool is made of hardened tool steel, and is operated in a hydraulic press. To provide a smooth finish on the outside of the cup, the inside of the upper tubular punch/die is highly polished, as the drawing operation squeezes the material against that face and will reproduce any imperfections.

Note that during the drawing operation, the sides will thin because the metal stretches to provide extra depth while maintaining a constant inner diameter. The bottom corners of the punch, and the corners at the entry to the die, should be rounded to prevent a tendency to shear the base instead of pushing it. Clearances of up to 15 per cent more than the thickness of the sheet should be allowed between punch and die, to allow for metal flow under pressure. Some experimentation will be required to determine the extent of the thinning, but it can amount to up to 25 per cent of the original thickness.

Finally, use plenty of oil to lubricate punch and die during drawing operations.

There are videos of this tool on the NYC CNC YouTube channel, and some additional details on the NYC CNC Facebook page.

Fig. 9-15: The copper strip is guided inside the upper part of the tool, for blanking. JOHN SANDERS AT NYC CNC

TURNING FLANGES

Punches, dies and formers are not simply for conventional blanking, punching or drawing, but can be used to persuade a workpiece to change shape, such as a rolled end on a tube. Proceed as follows:

Fig. 9-18: Flaring cone.

- Begin by using a conical former to flare the end of the tube (Figs 9-17 and 9-18).
- Then flatten the end of the tube (Fig. 9-19).
- Then force the tube through a die with an internal diameter (ID) = the outside diameter (OD) of the rolled end (Figs 9-20 and 9-21) to form the return bend and finish the roll (Fig 9-22).

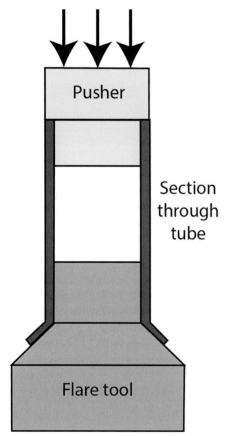

Fig. 9-17: Using a conical tool to form a flare on the end of a tube.

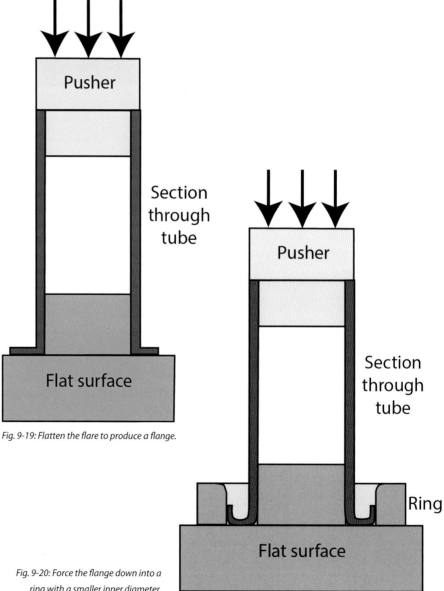

Fig. 9-19: Flatten the flare to produce a flange.

Fig. 9-20: Force the flange down into a ring with a smaller inner diameter.

Fig. 9-21: Pusher, tube and ring, with the upturned roll visible at the inner surface of the hole.

This kind of operation could be performed using a jenny and a pair of grooved formers, but only on larger, thinner tube. Both methods form a U-shaped burr or turn on the end of the tube.

Fig. 9-22: The finished roll.

10 Surface Finishing

No matter how well constructed an item may be, it is the surface finish that puts the icing on the cake. Two similar items might be equally well made, but they may be categorized by their surface finish: thus one may look better than the other, which might suit it for use in a public area where appearance matters, while the other, though equally functional, might be consigned to a workroom or utility area because it looks less attractive. Furthermore a well-chosen surface finish might protect an object, preventing it from degrading in service. Rust may be kept at bay, or corrosion due to sea water might be slowed, and for an object used outdoors, the worst effects of the weather might be minimized for a time.

The materials used to construct an object can determine its durability and sometimes its appeal, but even where appropriate materials have been chosen, the final finish can make a significant difference to its performance as well as its visual appeal.

CHOOSING YOUR MATERIAL

The first stage in finishing takes place before fabrication begins, and decisions made at the design stage can fundamentally affect the final finish that can be achieved. Making an item in steel as opposed to stainless steel, or in brass as opposed to aluminium, means that some finishes are possible to achieve, but others are difficult or impossible to realize. Both aluminium and stainless steel can be given a very high polish, for example, but while stainless steel will retain that finish outdoors for years, aluminium will dull, tarnish and corrode much more rapidly in the same environment. Design features such as narrow internal channels and sharp corners are difficult to finish properly, and difficult to protect from the effects of the atmosphere or the weather.

Sometimes the choice of materials is limited, as when making something designed to be used in the food industry. Here, stainless steel is a favourite because of its corrosion resistance, especially where water and air mix. But welds need to be located where they can be properly finished, otherwise they act as a focus for corrosion.

So design is the first consideration, and it enables or limits the choice of finish in many cases. Specifying the material finish desired involves consulting the material suppliers to find out what finishes are available. Stainless steel, for example, is commonly available in five finishes from the rolling mill, varying from non-reflective and rough, through dull, to a bright reflective finish (Fig. 10-1). A number of special finishes are also available from some suppliers, from a coarse ground finish through bright, polished and coloured to patterned and surface coated. Choice of the appropriate finish at this stage can save a great deal of work in getting a completed fabrication to an appropriate final appearance.

OPPOSITE PAGE:
Stainless steel can be given a very fine polish, and coloured by applying oil and heating.
RMS ENGINEERING LTD

Fig. 10-1: Stainless tube and sheet can be supplied in different finishes.

SURFACE PREPARATION

Unless your finished work is designed to have a rusty finish – such as a garden statue or the façade of a building made from Corten steel, which is intended to be left to rust – you will want to make an effort to remove rust, oxides, stains, scratches and other undesirable features from your finished work before applying a protective coating. Furthermore some materials require relatively little surface preparation, while others may require more extensive attention.

Most applied finishes, such as a paint or powder coating, must be able to grip the surface on which they lie, so although the surface needs to be well prepared, it is important that it is left sufficiently rough to provide a grip for the coating as it dries. That is not an excuse for leaving deep scratches and gouges, however, because the required roughness is barely perceptible to the human eye, and visible scratches are likely to remain visible under the coating, as well as detracting from the finish because they have relatively sharp edges to which the coating will be less inclined to stick.

The stages in preparing a metal surface to receive a coating are as follows:

- First eliminate all deep scratches, hammer marks, hills and hollows, and round all edges by deburring, using a file followed by abrasive paper (such as emery paper or wet and dry used as dry as possible).
- Clean the surface thoroughly by washing or wiping with a degreaser. Use clean cloths or paper, and preferably finish with a low-tack cloth then a lint-free cloth.

The stages required after this depend on the condition of the surface of the metal at this stage. Some rolled steel sheet, such as the commonly available 3mm sheet, has a black/blue oxide surface (Fig. 10-2) which

Fig. 10-2: Thicker steel sheet may have a blue oxidized finish.

is often rusty, and this needs to be stripped back to steel-coloured bare metal:

- A hand-held grinder (Fig. 10-3) is often the first tool used for stripping back a metal surface, but the grinding discs commonly supplied for removing metal are far too coarse for this task and leave many deep scratches, as well as ruining the flatness of the surface.
- One solution is to change the grade and type of disc, choosing instead 40-, then 60- or 80-grit Zirconium flap discs (Fig. 10-4): these will remove metal quickly, but leave a much better finish.
- Follow that with a polishing pad (Fig. 10-5) or emery paper, with increasingly

Fig. 10-4: Flap discs are available in a range of abrasives and grit sizes: the higher the grit, the finer the abrasive.

fine grit sizes of up to 400 for a very fine finish. Choosing the right grade of grit for the degree of cleaning required is important: too coarse a grit and the surface will be unnecessarily deeply marked, requiring a lot of subsequent work to reach a fine finish, while too fine a grit initially will result in much extra work attempting to produce a clean surface.

- When using a flap disc or a grinding disc, maintain a shallow angle between disc and sheet, with the disc

Fig. 10-3: A hand-held grinder fitted with a 115mm Zirconium flap disc.

Surface Finishing • 111

Fig. 10-5: Fine abrasive pads are usually softer and are impregnated with higher-numbered abrasive grits, from 100 to 400 and higher.

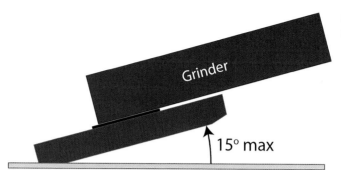

Fig. 10-6: The ideal angle of tilt for a grinder is 10–15 degrees.

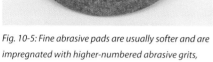

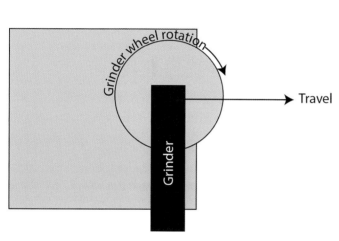

Fig. 10-7: Grind off the edge of a sheet to avoid catching the workpiece or rounding the edge.

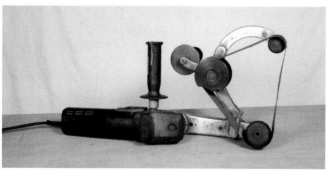

Fig. 10-8: A belt grinder for tubes and convex surfaces.

Fig. 10-9: Pressing the free section of the belt against the tube lowers the arms and wraps a section of the belt around part of the curved surface.

tilted at no more than 15 degrees (Fig. 10-6). Take care with the edges and make sure that the disc rotates off the edge and not into it (Fig. 10-7): rotating towards an edge will tend to catch the edge, which is dangerous, but will also rub and spoil its shape.

- Tubes pose a particular problem because of their curved surface, but variable speed belt grinders (Fig. 10-8) fitted with appropriate grades of belt simplify surface treatment (Fig. 10-9).

Surfaces can also be cleaned using shot, sand or bead blasting, with the type of abrasive chosen to match the degree of cleaning and the grade of final finish required. Acid dipping can clean steel panels and remove rust from steel parts, leaving them clean and ready for surface treatment. This is an industrial process, but not overly expensive given the work it can save.

SURFACE FINISHING METHODS

Painting

Most metal objects can be painted. Some take paint better than others, meaning that paint can get a better grip on the surface of some metals as compared to others, but there are some features of paint and painting that are common to most metals. Good design and attention to detail at the initial planning stage can make a difference to the ease with which paint can be applied, and its performance in service. Sharp edges and corners should be avoided if possible, because paint will tend to draw away from sharp edges and features where there is a sudden change in direction. Paint is a liquid, and behaves like any other liquid, so it flows and is subject to capillary action and the effects of surface tension, and these characteristics impose restrictions on its effectiveness on some surfaces.

Most paints require a primer, which is a first coating designed to bond to the metal surface, and to give a finish that will provide an effective key for paint. As a general rule, the protection provided by paint increases along with the thickness of the coating. A primer acts as a first coat, and some primers are specifically designed to provide protection against corrosion, with some of those containing zinc, for example. Some primers are also designed to be applied to a rusty surface, in which case they chemically neutralize the rust.

Some metals, notably brass, require an etching primer, which contains an acid. The acid etches the metal, providing a base to which the paint can adhere. Without the acid, the surface of the brass is too smooth for paint to grip it securely. Similarly zinc-coated metal sheet requires treatment with a specially designed primer that will etch the zinc, bond to it, and provide a good basis for paint. Phrases such as 'special metal primer' and 'direct-to-zinc' can be found on specialist paint tins.

Paint can be applied by brush, roller or spray, and the method chosen often depends on the size of the job. Large areas may be best sprayed, and a sprayed finish is often superior to the surface texture left by a brush or roller. One of the reasons is that paint is thinned before spraying, another is that a practised hand can often spray an even coating; also the speed at which the paint droplets hit the surface causes the paint to spread out, achieving good coverage as adjacent droplets merge.

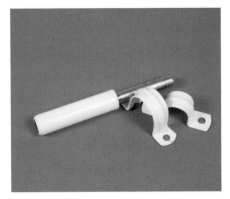

Fig. 10-10: Items that have been dip coated in a fluidizing tank.

Dip Coating

Dip coating is a simple technique for applying a plastic or synthetic rubber coating to an object (Fig. 10-10), and is the oldest of the dip coating processes. The coatings are supplied in several forms, including liquid (in a paint tin), aerosol and powder. The liquid and aerosol forms are liquid at room temperature while in their containers, and may be applied with a spray, brush or roller, then left to dry in the air. When dry, the coating provides a flexible, insulating, chemically resistant coating. When applying to metal, which is a non-absorbent surface, use an appropriate primer first. This process is simple, easy and effective, especially for small objects.

Dip coatings can also be supplied in powder form, and these are usually polyethylene ('polythene'), nylon, cellulose

Fig. 10-11: Fluidizing tank. Powder is available in a range of colours.

acetate butyrate (CAB, or 'butyrate') or polyvinyl chloride (PVC). Of these, PVC is often used commercially for coating the handles of tools, but PVC and CAB are not as readily available as polyethylene or nylon, especially in small quantities. Powder-based dips are applied using a fluidizing tank, in a process whereby air is passed through lightweight powder, causing the powder to float inside the tank like a fluid (Fig. 10-11).

In practice, when the air is turned on, the level of the powder rises slightly, but the key change is that the aerated powder offers little resistance to an object lowered into the tank. The work is heated and lowered into the tank, then withdrawn almost immediately. The heat causes the powder to melt and stick to the surface of the object, giving it a fairly uniform coating. As with paint, adjacent particles flow and merge, covering the surface.

The coating achieved with a fluidizing tank can be relatively thick, offering good protection once it has cooled. This process is easy to carry out, especially for small objects, and takes place in four stages, where the part to be coated is:

- cleaned and degreased
- heated (3 to 5 minutes in an oven at 250°C)
- dipped in the plastic in the fluidizing tank
- left to cool (usually hung up somewhere out of the way)

As the coating cools, it hardens, producing a hard-wearing and attractive surface.

Note that, at the heating stage, it is important that the heating method does not introduce dirt on to the surface of the work, so naked flames are not recommended for this task.

Liquid dips are easier and quicker to apply, but the fluidizing tank gives a more even coating. Liquid dip coatings can be built up by repeated dipping, but the use of a fluidizing tank is a one-off process because it would require reheating the work, and that would damage the initial coating.

In both cases this is a quick way of applying a durable surface in a range of colours at low cost. The coating is not as hard as some other types, but that can be an advantage, especially where the object is to be gripped.

Powder Coating

Note that the term 'powder coating' has a slightly different meaning to that implied in casual everyday use, and does not usually refer to the application of plastic dip coatings in powder form.

In the powder-coating process, powder is applied to a metal part electrostatically, by electrically grounding the work and applying an electrical charge to the powder, then spraying powder towards the work. The difference in electrical charge attracts the powder to the work, which is especially useful for tubing, where the powder is attracted 'around the curves', helping to coat the tube evenly and minimizing waste. The part and powder are heated, and the powder melts and flows rather like paint, so that adjacent particles merge to form a complete coating. The heat then cures the powder.

Powders are available in a range of colours, and the finish is usually shiny.

This is a popular finish because it is relatively durable, although that durability depends on the thickness of the powder and the environment in which it must protect the object. Motorcycle frames, car wheels and many consumer products are powder coated, as this is an industrial process suited to the mass coating of multiple items.

Powder coating can also be carried out in the workshop for one-off items, provided sensible health and safety procedures are followed, and the cost is not excessive. Follow the instructions provided with the equipment.

There are two types of powder coating gun: the corona and the tribo. The corona gun works by using a very high voltage to generate charged particles, which are fired at the powder as it travels along the nozzle of the gun. The powder is charged by the time it leaves the nozzle.

The tribo gun works by rubbing the powder along the inside of the nozzle, using friction to generate a charge in the particles. The composition of the powder and the material of the nozzle walls is different, and specially chosen so that a charge is generated by these dissimilar materials. As with the corona gun, the powder is charged by the time it leaves the nozzle, but the extent of the charge on the powder particles is not as great as with the corona gun. Wear is a factor too, and the nozzle of the tribo gun is subject to more wear due to friction between powder and nozzle walls.

Corona guns were first on the scene and are most readily available, but the tribo gun has much to commend it. The choice, however, will depend on the composition of the powder, and the tribo gun is not suitable for some specialist materials such as powdered porcelain enamel ('Frit'), as these are too abrasive.

Metal Coating

What better way to protect a metal workpiece than to coat it with another metal? Coating steel with zinc, for example, will provide an outer protective layer that protects the steel from moisture. If the coating is scratched, exposing the steel, the zinc corrodes sacrificially (that is, instead of the steel), providing further protection. Chrome or nickel plating both provide protection, and an attractive finish often seen in consumer goods or public places.

Zinc can be applied commercially using a hot dip process, where a cleaned steel item is dipped into a bath of molten zinc. It is not, however, a process that can easily be carried out in the smaller workshop. Instead, chemical/electrical plating methods can be used to create a layer of the protective

> **ELECTROLYTES**
>
> Ions are particles carrying a positive or negative charge. This causes them to be attracted to the terminal of a battery that has the opposite charge. Positive ions will be attracted to the negative pole of a battery, and negatively charged ions will be attracted to the positive terminal.
>
> An electrolyte is a substance which contains free ions. Because the ions are attracted to the opposite terminal of a battery, an electric current can flow in the electrolyte.
>
> As current flows, it transports ions across the electrolyte. In a plating solution, a block of metal (e.g. nickel, for nickel plating) is connected to the positive terminal of a battery, and the work to be plated is connected to the negative terminal. The plating metal and the work are suspended in an electrolyte fluid, so that current flows through the electrolyte, from the positive to the negative terminal (from the metal to the work). The current acts on the anode (the nickel, in this case) and effectively dissolves some of it. Those particles are transported through the electrolyte by the movement of the ions, and are deposited on the other terminal (the work). This leaves a coating of the anode metal on the work, so, in this case, the work is nickel plated.
>
> The key to all of this is the presence of ions in the electrolyte.

metal on the work. So steel can be plated with zinc, nickel, copper, brass, bronze or a form of chrome finish by using appropriate chemicals and a low-voltage power supply. Plating involves the following procedure:

- Cleaning and degreasing the work; this must be done chemically to a high standard.
- Suspending the work in a container full of dissolved chemicals that form an electrolyte, along with some pieces of the plating metal which act, electrically, as anodes, providing a supply of current through the fluid. The work is

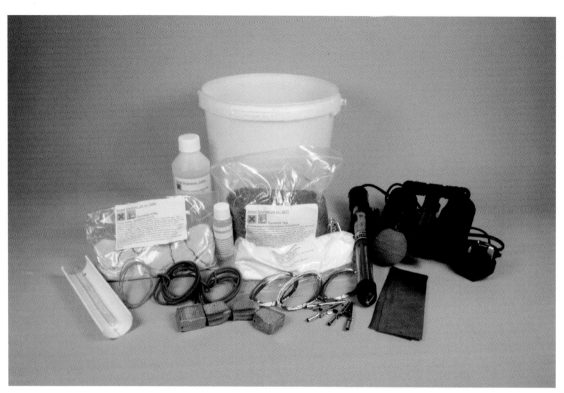

Fig. 10-12: Chemicals and equipment for nickel plating small items.

connected to the negative terminal of a power supply, and the anodes to the positive terminal. The power supply should have a reliable means of controlling the voltage and the current passing through the fluid, as these are critical to the success of the process.

- Once plating has been completed (which may take 20 to 60 minutes, depending on the thickness required), remove the work, disconnect it, and rinse it thoroughly. Some plating may then benefit from polishing.

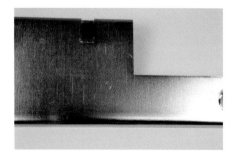

Fig. 10-13: Zinc-passivated steel has a characteristic yellow colour.

Fig. 10-12 shows a typical home plating set-up, and Fig. 10-13 a typical zinc-passivated finish on steel.

POLISHING

Some metals can be polished to a high shine, and look very attractive – and that shine does more than look good: it protects the surface of the metal. On its own, though, it will degrade over time and dull, and then allow the metal to corrode. Some metals, such as gold and silver, dull little, while others, such as bronze, develop an attractive patina as the outer surface corrodes and dulls, but without allowing the metal to corrode to any significant extent. Brass dulls and is prone to staining (Fig. 10-14), but can usually be revived quite easily by repolishing. Steel suffers badly, however, corroding quickly and extensively, and continuing to rust until it finally disappears. Stainless steel overcomes that problem, and because of its composition and its chromium content, it not only takes a brilliant shine (Fig. 10-15), but resists corrosion (although that does depend on the grade of stainless steel, and its environment).

So for some metals, polishing is sufficient, while for others, polishing should be followed by a lacquer, wax (Fig. 10-16) or some other protective layer, which might let the shine be seen while providing protection from the elements.

Fig. 10-14: Brass tarnishes and may mark badly, requiring the use of abrasives to remove the stains.

Surface Finishing • 115

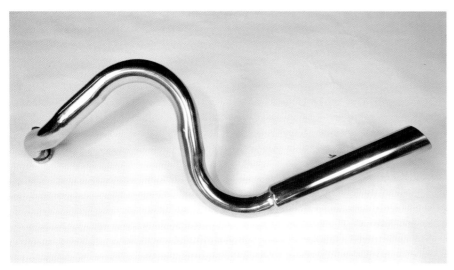

Fig. 10-15: Polishing can impart a fine shiny finish.

Fig. 10-16: Polished surfaces can be protected by applying a thin layer of microcrystalline wax, which is then polished to a high finish.

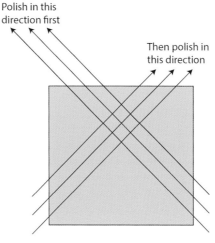

Fig. 10-17: When polishing using finer grits, alternate the polishing directions.

- Rub the abrasive on the surface, using mechanical means if possible, and using grinding and polishing discs, belts, rollers or mops (for the final stages)
- At each stage, polish in a different direction to the previous stage (Fig. 10-17) so that the existing scratches are not being deepened but are having their tops lowered. Each stage should use a finer grade of abrasive, and may progress from a disc to an abrasive roller and then to a mop
- You may choose to give the work a grained finish by polishing in one direction, using a roller instead of a mop, but that should be the final stage, and previous stages should be polished in alternating directions
- Final polishing should take place using a soft cloth

Some metals may be taken to a higher shine by adding another stage and polishing on a powered rotating buff (Fig. 10-18). These carry felt, cloth or rubber wheels which can be charged with abrasive paste (Fig. 10-19), usually by rubbing a bar of the abrasive across the edge of the wheel. Press the metal against the wheel, turning it carefully so that the wheel contacts each part of the work in turn.

Take extreme care with this kind of

Polishing consists of abrading the surface of the metal with successively finer media, until the surface is covered in tiny scratches, invisible to the naked eye. Those scratches reflect and scatter light, producing a shine.

The polishing process is simple, if repetitive:

- Begin with the finest abrasive, which will remove imperfections in the surface. These range from lumpy welds to deep scratches, both of which may require the use of a relatively coarse abrasive

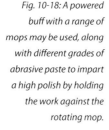

Fig. 10-18: A powered buff with a range of mops may be used, along with different grades of abrasive paste to impart a high polish by holding the work against the rotating mop.

Fig. 10-19: Polish for use on a powered buff is normally supplied in a bar, and is rubbed against the rotating mop, charging it with abrasive.

operation, as there is a real risk of severe personal injury. It is wise to ensure that the wheel rotates away from any edges (Fig. 10-20), because if the wheel catches an edge it will tear the work from your hands and fling it back at you. If you have read the introductory notes on Safety First, you will know that flying sheet metal and tube behave like giant razor blades, but are capable of doing much more damage. Try to stand slightly to one side of the work and wheel, and pay attention to your balance and your grip, keeping your hands away from the wheel.

Working through progressively finer grades of abrasive paste produces the most beautiful shine. Clean the work, then polish with a soft cloth or a clean rotating fleece.

Polishing Brass

Brass is best polished carefully, by hand, using abrasive paper wrapped round a sanding block. Particular care should be taken at the edges, as it is all too easy to round these. Brass plates can be butted up against a ledge or sheet of the same thickness, to help protect the edges.

Use progressively finer grades of paper, up to 1200 or 2000 grit emery, then progress to polishing papers in grits from 0 (coarsest) through 0/2 to 0/8 (finest), then crocus paper and finally polish (Brasso or Solvol Autosol). The Micro-Mesh system of flexible abrasive papers may also be used, in grades from 1500 (coarsest) to 12000 (finest): this system claims to be able to produce a surface finish with scratches no larger than twenty microns (Fig. 10-21). In the final stages, wear cotton gloves or finger cots (Fig. 10-22), and do not touch the surface of the metal.

Although laborious, this technique is used by clockmakers and can produce the very finest finish on brass.

Protect the finished surface with microcrystalline wax, applied carefully and thinly, and polished until it virtually disappears: beautiful.

Fig. 10-21: Micromesh abrasives are available in a range of very fine grit sizes, and can impart an extremely fine finish.

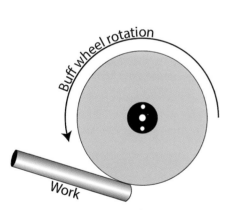

Fig. 10-20: A rotating mop must always polish away from an edge and away from the operator.

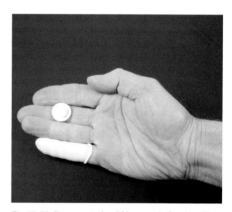

Fig. 10-22: Finger cots should be worn before handling polished work.

Projects

Project 1: Fuel Tank for a Model Aircraft

Fig. P1-1: Fuel tank for a model aircraft.

Model aircraft fuel tanks can be made of tinplate and copper pipe, using soldered joints (Fig. P1-1). A typical tank is shaped to fit inside the fuselage, near the engine. On a Control Line model which flies in a circle, the tank will be shaped so that the fuel is directed towards the feed pipe as it is thrown against the outside of the tank by centrifugal force.

There are normally three pipes: a fuel delivery pipe to take fuel to the engine, a filler pipe, which is capped when the tank has been filled, and a vent pipe which overflows when the tank is full. The vent pipe is left uncapped and faces into the airstream so that the tank is pressurized as the aircraft flies.

THE DESIGN

Fig. P1-2 shows the two tinplate sheets required, and Fig. P1-3 shows the pipes. Fig. P1-4 shows the orientation of the fuel delivery pipe within the tank. It does not quite touch the opposite side. Inside the tank, the bottom ends of the filler and vent pipes do not touch the opposite surfaces, but sit

Project 1: Fuel Tank for a Model Aircraft • 119

Fuel tank: Flat patterns
Material: 0.2mm tinplate

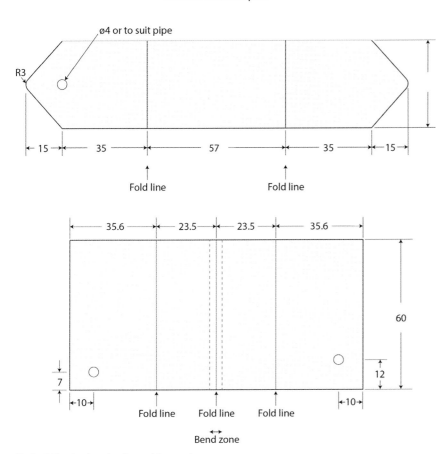

Fig. P1-2: Flat developed surfaces of the two sheet parts.

Fuel tank pipes
Material: copper pipe ø4mm OD

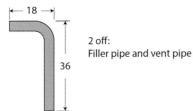

2 off:
Filler pipe and vent pipe

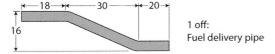

1 off:
Fuel delivery pipe

Fig. P1-3: Dimensions of the fuel tank pipes.

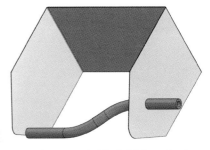

Fig. P1-4: The inner end of the fuel delivery pipe sits clear of the opposite side.

Fig. P1-5: The filler and vent pipes are suspended 1 to 2mm above the opposite surfaces to allow fuel into and air out of the tank.

about 1–2mm clear, to allow fuel into, and air out of the tank (Fig. P1-5).

MATERIALS

- Tinplate sheet 0.2mm × 150 × 100
- Copper pipe; 4mm OD, 2.7mm ID; approximately 250mm long. Note that this kind of soft copper item is often called a tube, so check the actual external and internal diameters carefully before purchase.

Preparing the Materials

Using a scriber, a ruler and an engineer's square, mark the shape of the two sheet parts, including lines to represent the folds. The flat developed shape includes an allowance for bends, but where the dimensions

Fig. P1-6: Sheet parts for the tank.

make it difficult to mark accurately, just round up the sizes to the nearest millimetre. Alternatively, cut a paper pattern and stick it to the tinplate.

Mark the positions of the holes, and centre pop each one. A pricker punch is enough for this job.

Using sharp snips with close-fitting blades, cut out the two main parts (Fig. P1-6). Small tinsmith's or jeweller's shears work better than large shears intended for thicker sheet metal.

MAKING THE FUEL TANK

Forming the Holes

Use a small drill (1mm or 0.040in) to drill a pilot in the centre of each hole. Hold each drilled hole centrally over a larger hole drilled through a block of metal. Then use a centre punch or a tapered punch to open out the hole until the pipe can just pass through. This creates a sleeve for each pipe, to help strengthen the soldered joints.

Fig. P1-7 shows a punch of the same diameter as the pipe and with a long slow taper on the nose. The punch is held in a staking tool, centred above a die. The diameter of the hole in the die is the punch diameter plus twice the thickness of the material, and it has a chamfer at the entry (Fig. P1-8).

The centre of the hole is marked with a pricker punch, then drilled through with a

Fig. P1-7: Piercing punch in a staking tool.

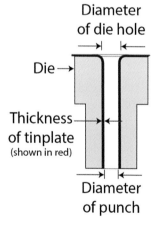

Fig. P1-8: Dimensions of the die that accompanies the piercing punch.

1mm drill. The material is positioned over the die, and centred by gently lowering the punch so that it sits in the 1mm pilot hole. Tap the punch sharply with a hammer, and continue tapping until the main body of the punch passes through the hole.

HARDENING AND TEMPERING THE PUNCH AND DIE

The punch and die are made from silver steel (high-carbon steel). For use in thin material such as tinplate, the punch and die do not need to be hardened, but for multiple use, or when working with thicker material, both punch and die should be hardened, then tempered to a pale straw colour. Follow the instructions given by the manufacturer of the steel used for the punch and die. When using silver steel, heat to a bright cherry red and, in the case of the die, hold at that temperature to allow the material to heat right through.

Quench in water, taking great care to drop the punch vertically into the water. There is a real danger that a punch of such a small diameter but relatively great length will warp as the water cools it rapidly. Test the hardness by trying to file the surface of punch and die: if they are hard, the file should slip rather than bite into the material. Clean and polish the punch and die.

Then heat until the working surfaces (the tip of the punch, and the top face of the die) show a pale straw colour, and quench immediately in water. Clean and polish ready for use.

Folding the Two Main Sheet Parts

On the side/rear/side piece, using a small folder, or a vice with plain jaws, grip the rear (centre) section and fold the sides (Fig. P1-9) so that the punched sleeve faces inwards. Using a vice, the second fold may foul the vice behind the top jaw; it depends on the design of the vice.

On the top/front/bottom piece, start with the top and bottom bends and make those by comparing them with the angle of the mating sides (Fig. P1-10). Both punched sleeves should face inwards. Then make the small bend in the centre of the sheet, forming this around a rod (Fig. P1-11).

Project 1: Fuel Tank for a Model Aircraft • 121

Fig. P1-9: Folding the sides and rear of the tank.

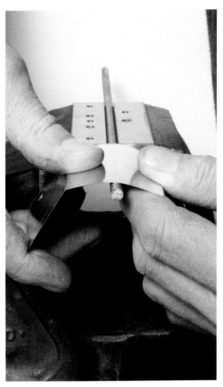

Fig. P1-11: Bending the nose round a rod.

Bending and Cutting the Pipes

Using a small tube bender, or an external spring and finger pressure, bend a length of

Fig. P1-13: Bending a copper pipe supported by an external bending spring.

pipe to shape (Figs P1-12 and 13). The pipe can remain attached to the uncut length for this operation, and be cut to length after the bend has been made.

Create a template drawing for each pipe, on a sheet of paper, and use that as a guide when bending. The simplest drawing, for the filler pipe, would be two lines at right angles, with marks showing the beginning and end of the pipe (Fig. P1-14).

The drawing for the fuel delivery pipe should show the position of each bend, and the offset between the two parallel sections

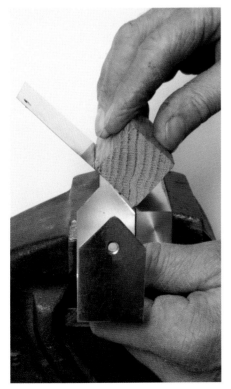

Fig. P1-10: Making the bottom/front fold to match the angle of the taper on the sides.

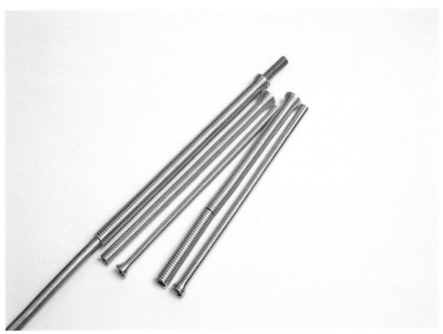

Fig. P1-12: External bending springs to support small-diameter pipe during bending.

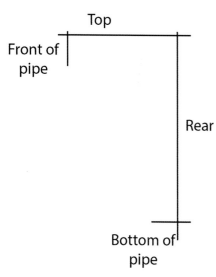

Fig. P1-14: Filler tube template drawing.

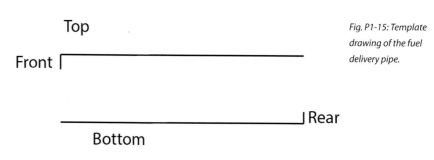

Fig. P1-15: Template drawing of the fuel delivery pipe.

of the pipe (Fig. P1-15), but a full drawing might be easier to follow. Lay the pipe on the drawing (Fig. P1-16), mark it off, and cut with a fine saw. Repeat for the other pipes.

Joining the Parts

There is a choice, here: you can solder the main parts together, then solder the pipes into the tank (*see* below); or you can solder the pipes into the holes, then solder the sheet parts together.

Soldering the Sheet Parts

Clean the tinplate, especially where there will be soldered joints.

Using flux, then solder, tin the joints on the inside and outside of the joint lines, and along the top and bottom of each vertical side (Fig. P1-17). If you want to use tabs or angle pieces to support the joints, sweat them in place on the inside of the joint positions on the larger top/front/bottom piece.

Place the two main parts in their final positions, and solder together, running a little fillet of solder along each joint (Fig. P1-18).

Soldering the Pipes

Clean the outside surfaces of the pipes, then flux and tin at the positions where you will make the soldered joints (Fig. P1-19).

Insert each pipe into its hole, taking care with the depth, and solder in position by applying the flat part of the chisel point, and most of the heat, to the rod, while the narrow tip of the iron is against the tank. Apply solder to the opposite side of the pipe at the joint when it is hot enough.

When positioning the pipe, a 'third hand' device with stand and clamps is useful to hold the pipe upright, facing forwards, and with the inner end at the correct depth inside the tank.

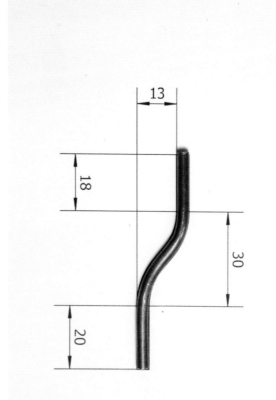

Fig. P1-16: Template drawing showing the bend positions and a drawing of the pipe (underneath the copper pipe).

Fig. P1-17: Tinning the inside edge positions.

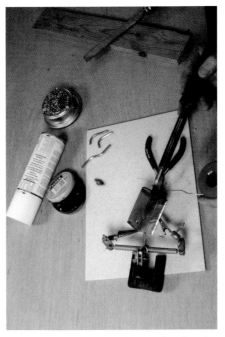
Fig. P1-18: Soldering the seam between the side and the bottom of the tank.

Fig. P1-19: Tinned pipe.

FINISHING OFF

Clean off any excess flux using flux cleaner, then tidy up the corners and edges with a file to complete the tank.

Cut a strip of tinplate and bend it to fit around the tank, leaving enough for a tab at each end (Fig. P1-20). That can be used to secure the tank in position on the model. Mark, then drill a hole at each end for a securing screw.

Once the tank is on your model, fill it with fuel, swing the prop, and enjoy performing impressive aerobatics to amaze the envious bystanders!

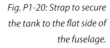
Fig. P1-20: Strap to secure the tank to the flat side of the fuselage.

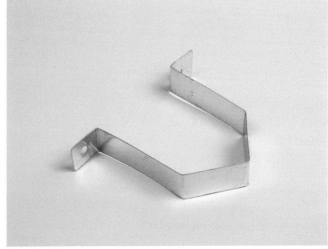

Project 2: Motorcycle Ammeter in a Tin Can

Fig. P2-1: The main ingredients are a can and an ammeter.

An ammeter is a simple device which monitors the current in the battery-charging circuit. This one is designed to be mounted on a motorcycle, but it could just as easily be used in a car. Ammeters are often mounted in a circular hole in the dashboard of a car or the headlamp shell of a motorcycle; this one is in an old tin can (Fig. P2-1).

THE DESIGN

The ammeter has a set of four tabs attached to the bezel to hold it in position, but they also serve to locate the ammeter in a hole in a panel (Fig. P2-2). The hole on the end of the can needs to be just large enough to allow the ammeter to enter, but it should be a firm fit. Then, when the tabs are spread a little inside the can, they will hold it in position.

There are two ways to do this: the first is to cut a hole large enough for the ammeter to slip in at any position; the second is to make the hole a close fit on the body of the ammeter, and to add four slots where the tabs will be. The second method helps prevent the bezel and meter from turning due to vibration. Fig. P2-3 shows the difference.

Fig. P2-2: The ammeter has four tabs under the bezel.

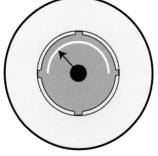

Fig. P2-3: Two ways of using the tabs to secure the ammeter.

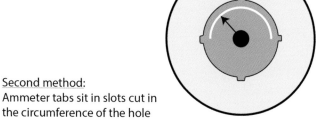

First method:
Ammeter tabs sit against the circumference of the hole

Second method:
Ammeter tabs sit in slots cut in the circumference of the hole

Project 2: Motorcycle Ammeter in a Tin Can • 125

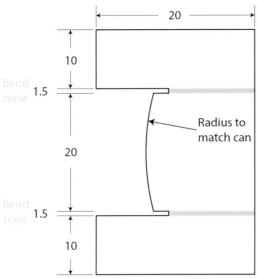

Fig. P2-4: Details of the internal brackets to secure the lid.

- M3 × 10mm cheese head set screws
- Small spring (two off) – optional
- Grommet to suit ammeter cables

Preparing the Materials

Open the can and enjoy the contents. Clean thoroughly, removing the wrapper from the outside surface. Any small bashes in the tin can be smoothed out by pressing from the inside with a dowel (Fig. P2-5), or by careful use of a small hammer with a domed face. Support the outside as you do this.

MAKING THE AMMETER CAN

Forming the Hole for the Ammeter

Stick a label across the end of the tin. Set a pair of compasses to a little more than half the diameter of the tin, hold the point of the compasses against the inside of the lip on the tin, and scribe arcs in four positions (Fig P2-6). The arcs will overlap, to give two lozenge shapes.

On each lozenge, join the end points. The centre of the end of the tin is where the straight lines cross (Fig. P2-7). Support the inside of the end of the tin on a dowel, and use a centre punch to mark the centre of the circle. Remove the label, set a pair of dividers to the radius of the hole you need, and scribe the circumference clearly (Fig. P2-8).

MATERIALS

- Tin can to suit the ammeter; the can should have a pull-tab opening top
- Mild steel sheet 0.68mm (or to suit – see text)
- Mild steel sheet 1.2mm
- 3mm (⅛in) short pop rivets (six off)
- M3 threaded rivets (two off)

Fig. P2-5: Bumps can be pressed out and smoothed from inside, using a dowel.

Fig. P2-6: Scribing arcs using a compass.

Fig. P2-7: Joining the end points with straight lines to find the centre point.

Fig. P2-8: Scribing the circle to be cut for the ammeter, using dividers.

Fig. P2-11: The ammeter sits in the hole. Tab positions are marked in red.

Fig. P2-9: Drill a hole in the centre of the circle.

The hole for the ammeter is relatively large, but the material is thin, so a hole saw will snag at it drills through. Simply drill a hole in the centre (Fig. P2-9), then use the very tip of a set of curved aircraft snips or a pair of small, strong scissors to cut a spiral from the drilled hole to the outside of the circle (Fig. P2-10).

The edge can be smoothed using a very fine half-round file (a fine 'Swiss' file, for example), filing across the edge at an angle rather than vertically.

If you have chosen to use notches in the circumference of the hole, file those now, using a very fine, small square or flat file. The ammeter should just slip into the hole (Fig. P2-11), but don't secure the tabs yet.

Making the Lid

Cut a circle of 1.2mm steel to fit neatly inside the open end of the can (Fig. P2-12). Mark the centre of that circle with a centre punch, and scribe a diameter; those marks will be hidden on the inside of the finished work. Smooth the circumference with a fine file.

Scribe a circle to indicate the positions of the two securing screws. Where this crosses the diameter, centre punch, then drill two 3.2mm holes for those screws.

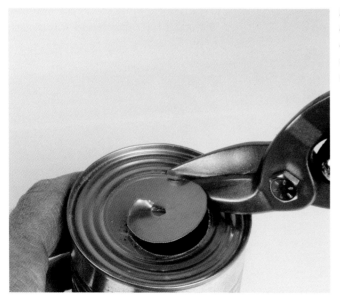

Fig. P2-10: Use the snips to cut in a spiral to reach the circumference of the circle, then cut round to create the hole for the ammeter.

Fig. P2-12: Cut a circle for the lid.

Dealing with the Grommet

Somewhere on the can, in a position to suit the design of the can and the way it will be mounted on the motorcycle, there should be a hole for a grommet to carry the cables to the ammeter. The can shown in the photos has this in the centre of the underside, but the position can easily be varied to suit. There are two support plates for the grommet hole, one on the inside and the other on the outside of the can. The diameter of the hole, and the thickness of the metal plates, should be chosen to suit the grommet.

On a larger piece of sheet, mark a strip to form the inner and outer grommet plates. Mark the centre of both plates, then mark the positions of the attachment rivets on one plate only (the outer plate).

Drill the rivet holes (in the outer plate only) with a normal twist drill, but use a butterfly drill for the larger grommet hole in each plate.

Cut the plates from the sheet, and finish the edges with a fine file. Round the corners with a small radius, by filing along the side and round the corner (Fig. P2-13).

Fig. P2-14: Grommet plates are curved against a tube of suitable diameter.

Fig. P2-13: Grommet plates.

Curve both plates by pressing them against something of approximately the same diameter as the tin (Fig. P2-14). The tin itself does not have sufficient strength for this. Because of the larger grommet hole, the plates will tend to fold sharply across the centre, so use a wooden mallet to tap the plates against a suitable curved surface.

Hold the outer plate in position against the can, and use a scriber against the inside of the grommet hole to mark the position of the corresponding hole on the tin. Drill a small hole (3mm or so) approximately in the centre, then use a conical grinding stone in a multi-tool to grind the waste material and finish the edges of the hole (Fig. P2-15).

Go carefully, and use a light touch. Ideally, the hole should be marginally larger than the hole in the outer plate, so that the sharp edges do not touch the grommet.

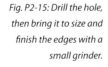

Fig. P2-15: Drill the hole, then bring it to size and finish the edges with a small grinder.

Hold the outer plate in position a second time, and mark, then drill the clearance holes for the two pop rivets.

Hold the inner plate in position, aligning the grommet holes, then mark the positions of the rivet holes. Remove the plate, centre punch the holes, and drill those holes. Support the plate from underneath while drilling, and either clamp it or hold it in a pair of vice grips. Go gently. You may have to re-curve the plate afterwards.

Carefully deburr all the holes so that the plates sit tightly against the tin. Then insert both rivets, and use a rivet gun to close them (Fig. P2-16).

Fig. P2-16: Rivet the grommet plates in position.

Making the Brackets for the Lid

The two securing brackets for the lid are made by folding. They are then riveted in position, leaving a small gap between the underside of the end plate and the brackets. This ensures that the end plate is drawn tightly against the end lip of the tin, and allows for a small spring on each set screw to counter any vibration from the engine which might tend to shake the lid loose.

Mark out the brackets. The sizes shown include an allowance of 0.5mm for folding. The bending allowance is actually 0.6mm, but that is too difficult to mark out from a ruler, and the difference will not affect the look or the function of the bracket.

Cut out the basic rectangular shape of each bracket, ignoring the recess for the curved end. To keep the edges flat, the brackets can be cut using a fine-toothed hacksaw: 32tpi works well for this thickness of steel.

To remove the shaped recess, make the two straight cuts as shown in Fig. P2-17. These cuts are extended a little beyond where they meet the inner curve, to avoid tearing during folding.

Fig. P2-18 shows that folding the tabs in a conventional folder will catch the tip of the curve. There is not enough space at the end of the folder to allow the arms to be folded one at a time, so they need to be folded individually in a vice, producing the result shown in Fig. P2-19.

The two sides of the bracket are now folded so that the tabs fold away from each other. This can be done in the side of a folder. One side is bent up by clamping on the body (Fig. P2-20). For the other side, the procedure must be reversed, clamping the side and folding the body up (Fig. P2-21). Think carefully about the clamping point and the thickness of the material.

To curve the tabs (Fig. P2-22), hold the bracket in position inside the tin, support the outside of the tin, and press the bracket firmly against the inside.

Decide where the lid-securing screws should be positioned in relation to the grommet hole, and mark the positions of the rivets for the brackets on the tin. Drill those holes.

Hold each bracket in position and mark the positions of the rivet holes on the brackets. Drill those holes carefully, then deburr all the holes. Rivet the brackets in position, making sure they are pushed fully home so that the curve at the back touches

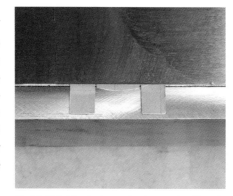
Fig. P2-18: Folding both ears simultaneously will catch the curved section.

Fig. P2-19: Fold the ears individually, in a vice.

Fig. P2-17: Lid bracket, showing the inner side cuts extending beyond the intersection with the curved section.

Fig. P2-20: Fold one side using the channel at the end of the folder to accommodate the folded ear.

Fig. P2-21: Working at the edge of a finger, fold the body up towards the ear and side.

Making a Fake Ring-pull Tab

To make the ammeter housing more impressive, make a fake ring-pull tab to sit under one of the securing screws on the lid.

Mark out the shape on a piece of sheet, leaving as much material as possible at the sides. Drill a hole to suit the bolt through a 16mm (or ⅝in) Q-Max cutter. If you don't have a butterfly drill of a suitable size, clamp the work firmly, then drill a small pilot hole first. Open up to final size using a normal twist drill, feeding gently downwards.

Finish the hole using the Q-Max cutter (Fig. P2-25). This has the interesting bonus that the edges of that hole will be curved inwards, on one face, rather like the edges on a real ring-pull tab.

Drill the small 3mm attachment hole,

Fig. P2-22: Pressing the bracket in position, to bend the ears, completes the shape.

Fig. P2-23: Rivet the brackets in position.

Fig. P2-25: Mark out for the ring-pull tab, then use a Q-Max cutter to punch the larger hole.

the inside of the tin, to give extra support against flexing (Fig. P2-23).

Put the lid on, and mark through the securing holes on to the brackets.

Very gently drill the brackets with a small diameter drill (around 1mm diameter), then open up to suit the M3 threaded rivet inserts. Fit those inserts (Fig. P2-24) and check the lid will be held tightly in place.

If you can find, or make, two soft springs to suit, these can be slid over the set screws and compressed between the underside of the lid and the brackets. They are not essential, however.

Fig. P2-24: Fit a rivnut to each bracket.

Fig. P2-26: The tab is finished except for a small fold.

Fig. P2 27: Mask the ends in preparation for painting.

Fig. P2-28: A coat of paint makes all the difference.

then cut the shape out and finish the edges by rounding with a file. Give them a radius like a real tab (Fig. P2-26).

Put a small fold just above the large hole, to subtly draw attention to the tab when it is fixed in position. It's the small details that make the whole project visually interesting.

Making the Securing Brackets

Securing brackets need to be designed to suit the position of the tin on the motorcycle, and the adjacent parts of the frame to which it will be mounted. They should provide a secure mount, but might include a layer of rubber between the frame and the mounting bracket, to minimize the transmission of vibration. The original ammeters were often mounted directly into a headlamp shell or the side of a battery box, without any concessions to protection against rattling, shaking or bumping – but that's no reason not to look after what is, essentially, a precision instrument.

Securing brackets may need additional fixing points on the tin, and those are probably best created before applying a final finish.

FINISHING OFF

Clean the can; mask the end rims and faces (Fig. P2-27), and use your choice of finish on the body (Fig. P2-28). Paint works well if the tinplate is first rubbed with steel wool to give the paint a key – but it is your choice.

Finish the lid and tab (Fig. P2-29). These could be painted, perhaps with an aluminium paint, but might look better nickel plated and polished – whatever you like.

Finally, mount the ammeter with its

Fig. P2-29: Can lid and ring pull tab.

cables attached (Fig. P2-30), not forgetting the O-ring (if there is one) which sits under the bezel, and bend the tabs to hold it in position. You might find space inside the tin to hold a polishing cloth as well, so that you can give your pride and joy a little shine during a pit stop.

Fig. P2-30: The ammeter fitted into the painted can.

Project 3: Fluidizing Tank

Dip coating metal objects is quick and easy if you have a fluidizing tank (Figs P3-1 and P3-2) to hold the powder and to blow air gently through the tank to aerate the powder and make it more fluid. The requirements are for low-pressure air, such as might be supplied by the 'blow' end of an old vacuum cleaner, or a small compressor or a fan.

Having a tank gives you access to the coating process and means your projects can look very professional and smart in a range of coloured coatings, comfortable to the touch.

THE DESIGN

Fig. P3-3 shows the important sizes for the tank. There is nothing critical about the sizes or the shape of the parts of the tank, and they can be made to suit the materials you have to hand, or the shape of the items to be coated. You could even make the tank rectangular.

MATERIALS

Sheet steel for the main parts:
- 1 piece approximately 1mm thick × 300 × 100mm
- 1 piece approximately 1mm thick × 300 × 700mm
- Steel strip for the valve, 3mm thick × 75 × 100mm
- Steel strip for the handle, 3mm or 6mm thick, 13mm wide and approximately 250mm long
- Cotton sheet at least 200 × 200mm

Fig. P3-1: Fluidizing tank.

Fig. P3-2: Fluidizing tank components.

132 • Project 3: Fluidizing Tank

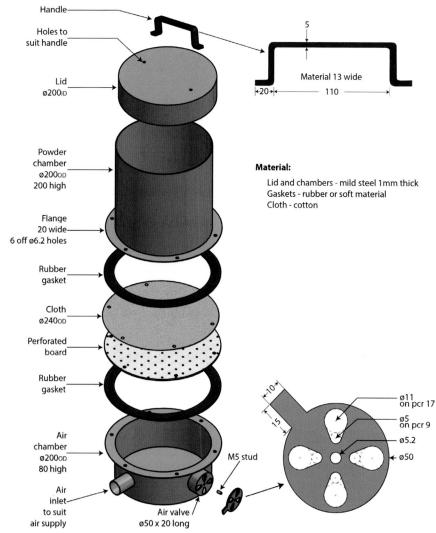

Fig. P3-3: Fluidizing tank sizes.

Material:
Lid and chambers - mild steel 1mm thick
Gaskets - rubber or soft material
Cloth - cotton

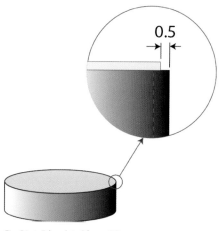

Fig. P3-4: Edge detail for welding.

- Rubber, or soft foam, or plastic, or cork for the gasket 250 × 250mm
- Perforated hardboard 250 × 250mm

MAKING THE TANK

Preparing the Sheet Parts

Calculate the width of the sheet required for the powder chamber, and cut to size. The width might include an allowance for a joggled spot-welded seam, or may be designed for a simple butt weld along the seam.

Calculate the width of the strip required for the side of the top lid. Cut this a little over-length, for final trimming to fit the completed powder chamber. A joggled joint won't work here, unless the overlap is on the outside.

Cut the circular piece for the top of the lid. You may wish to reduce the outside diameter by the thickness of the vertical sides, for welding (Fig. P3-4).

Cut the circular flanges, allowing for a small overlap on the inside, for welding to the powder chamber and the air chamber. Clamp the flanges together, and drill the holes for the securing bolts.

Cut the circular bottom of the air chamber, and the strip for the side.

Cut the perforated hardboard so that it matches the outer diameter of the flanges.

Mark out the shape of the air-valve plates, but do not separate them from the sheet at this stage.

Drill the holes to begin forming the petal-shaped cut-outs, and the centre holes. Then use a coping saw and/or a file to finish the cut-outs.

Finally, cut out the shape of each valve.

An alternative approach is to change the design of the air valve and the size of the tube to which they are attached, to make manufacture easier, as follows:

- Omit the tab of the outer valve plate. This can be made separately and welded on later
- If you have a large hole saw, make the outer diameter of the valve plates the same as the diameter cut by the inside of the hole saw (the bit that is normally the waste). The centre hole will have to be the diameter of the pilot drill for the hole saw, so that will be 6mm (or ¼in)
- You can also simplify the shape of the petal cut-outs, making them 13mm (or ½in) round holes

Make the air valve attachment tube by drawing the developed shape of the curve where the tube fits the chamber, then cut out the shape as a strip to be rolled into a tube.

The tube for attaching the air supply can be made to suit the hose supplying the air, or it could be fitted with a blank end and drilled to take a hose attachment. This tube can only be rolled on small-diameter rollers, but could be a short piece of ready-made tube. As with the tube for the air valve, the end of the tube should be shaped to fit the air chamber.

Make the holes in the sheet for the sides of the air chamber. This can be done by drawing the developed shape of the holes, which will be oval and not circular, in the flat. It could also be done later, when the sheet has been formed into the circular side, by placing the shaped end of the tubes against the chamber, marking the edges using a scriber, then carefully removing the centre of the hole, leaving a small allowance for welding. Or the hole could be drilled using a sheet metal hole saw with fine pitch teeth – with care.

The Powder Chamber and the Lid

Roll the cylinder for the powder chamber and weld the seam. Tack weld first, then stitch weld in stages to avoid distortion. Dress the weld, then lightly planish the join, over a curved former such as a large tube, or using a dolly.

Tack weld the base flange to the cylinder, then complete the weld in stages, making sure the flange remains flat. Do this as you go along, to minimize the work required afterwards.

Roll the side of the lid, then weld carefully, checking that the lid will be a close fit over the cylinder. Dress the weld on the inside, if necessary, then weld the top on to that short cylinder. This is best done with the short cylinder in position over the top of the powder chamber cylinder, but not fully down (to avoid welding the top to the powder chamber).

The Air Chamber

Roll the air chamber cylinder and weld closed, then attach the flange and the bottom end piece.

Weld the tubes for the air exit valve and the air supply to the wall of the air chamber.

Weld the threaded rod to one valve plate, by passing the rod through the centre hole and welding from the rear. Note that this hole may be threaded, if the original design is followed, which will help steady the rod during welding.

Then weld the plate to the edge of the supply tube.

If you made the operating tab separately, weld it to the other valve plate now.

Making the Membrane

Glue the cotton cloth to the perforated hardboard, applying glue only along a strip around the edge.

Trim when dry. Then use the flange as a template to mark out and drill matching holes for the securing bolts.

Making the Handle

The handle is best bent in a flat strip bender, or in a vice. Use heat if necessary, to make bending easier. The handle could be made thinner (3mm, for example).

Drill two attachment holes after bending.

Use the completed handle as a template to drill the matching attachment holes in the lid.

Making the Gasket

The gasket can be cut to shape, using a flange as a guide. Because the available material was not large enough to cut complete circles, the original gaskets were each cut in two parts, then glued together.

Because the gasket will be relatively soft, the holes for the securing bolts are best cut using a hole punch. This could be a commercially available punch, or it could be made by sharpening the edge of a convenient tube. The size is not critical.

When sharpening a tube, chamfer the outside of the edge, to leave the inner edge as the cutting edge. That way, the cut material will pass up the inside of the tube. Support the gasket on a soft wooden board which will not blunt the hole punch. This avoids having to harden the punch.

USING THE TANK

Load the powder chamber to half depth.

Open the air valve fully (the air exit valve, with the petal-shaped holes).

Connect the air supply at low pressure and volume if possible.

Gradually close the air valve until the powder has risen noticeably, and make a note of the position. Then turn off the air supply and watch the powder settle back down.

The work should be heated in an oven at 250–400°C. It will need an attached wire or strap which can be used to lower the work into the powder; this can be attached before it is put in the oven, or later, provided there is a suitable attachment point and the wire can be hooked on quickly.

The time required to bring the work to a suitable temperature will depend on its mass and whether there are any relatively thick sections.

Once the work has been heated, turn on the air supply and lower the work slowly into the fluid, then pull it back out. Hang it up and let it cool, and admire your professional handiwork.

Project 4: Spigot for a Workshop Dust Extractor

Fig. P4-1: Workshop vacuum extractor, showing the spigot for the hose.

A vacuum cleaner is a useful item in a workshop, especially if there is woodworking machinery which might create dust and chips (Fig. P4-1). A large hose connects to the vacuum, and attaches to machinery or is suspended near the source of dust.

The vacuum cleaner in this project has a cylindrical body, and a hose, fitted with an end cuff, connects to the vacuum via a spigot. Unfortunately the spigot is plastic, and the weight of the hose and its coupling sleeve has broken it (Fig. P4-2). This seems to be a common problem, and spare parts are available, but expensive. Rather than replace the spigot with an identical plastic part, a spigot can be made from metal. However, the standard sizes of metal tubes are different from plastic tube sizes, so this project involves rolling tubes as well as creating a flange for the spigot.

Fig. P4-2: The original spigot.

THE DESIGN

The replacement spigot consists of the external tube with flange, and the internal sleeve. The spigot must allow the existing hose cuff to slip inside, and should provide a good fit to avoid major leaks. There must be a stop inside the spigot, to stop the cuff entering too deeply. The rear of the cuff has a gentle swelling, and deep entry would tend to cause the swelling to jam in the spigot. The inside diameter of the air passage through the spigot should not be smaller than the passage through the cuff.

The external tube is of 3mm steel and has an internal diameter of 112mm, to match the external diameter of the sleeve. The tube will be made to match the hose cuff, and the other parts then made to match the tube. The outer end of the tube is plain, but the flange end must be shaped to follow the curve of the vacuum cylinder.

The flange attaches to the end of the external tube. The flange must be curved to match the external diameter of the vacuum body, and when the flange is bent, the hole to fit the external tube must be circular. This means it will be oval when the flange is flat, before bending. The same will be true for the external profile of the flange.

The inner tube must be a close fit in the external tube, but it needs a small flange or burr to act as a stop for the cuff. Inside the vacuum cylinder there are two bags: a plastic one with a hole which slips over the

Project 4: Spigot for a Workshop Dust Extractor • 135

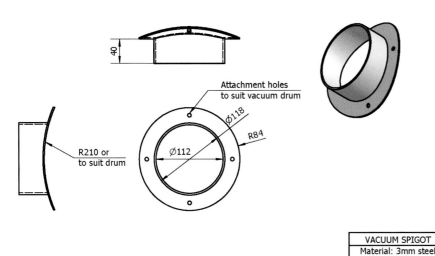

Fig. P4-3: Dimensions of the vacuum spigot.

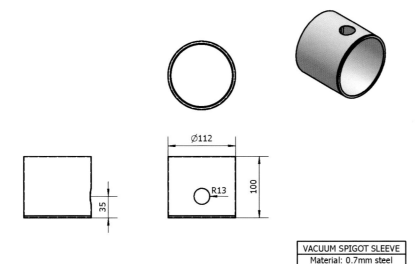

Fig. P4-4: Dimensions of the inner sleeve.

inner tube and is secured with an elastic band around the tube, and a fabric filter bag. The far end of the inner tube needs to be rounded to prevent snagging on those bags. Although the original plastic part did not have this feature, it would be useful to have a bead around the inner tube, to help retain the elastic band.

The inner tube also needs a hole to take a plastic pipe which feeds a little air into the space under the bags. This hole has a generous diameter and is a loose fit for the pipe, but it would be useful to have a sleeve round the inside of this hole, to prevent the sharp edge of the hole cutting or scuffing the plastic pipe.

Figs P4-3 and P4-4 show the spigot and the internal sleeve. Although the sleeve shows a flange at both ends, the free end could be plain and protected with panel edging tape, or it could have a rolled and wired edge, as shown in some of the photographs later.

MATERIALS

- Steel sheet 3mm thick: 1 piece 275 × 275mm and 1 piece 50 × 400mm
- Steel sheet 0.7mm thick × 130 × 350mm
- 500mm length of panel edging tape, or 100mm copper tube (optional)
- 500mm length of 3mm wire (optional)

MAKING THE SPIGOT

Drawing the Parts

If the developed surfaces of the tubes are designed using CAD software they will be too large to fit on A4 paper. The sizes could be taken off the screen, or the file could be printed in a print shop which has a plotter. Alternatively, the parts can be drawn on a large piece of paper, or on the workshop floor. The important sizes are for the development of the curve on the rear of the external spigot, and for the inner and outer shape of the flange.

For this project, the parts were drawn on a large piece of paper from a 2m long roll, and the sizes were transferred from the paper to the work using dividers and a ruler.

Mark the outlines using a scriber, then make punch marks along the lines using a pricker punch (Figs P4-5 and P4-6). This makes it much easier to see the lines as the shapes are cut out later. Make a punch mark every 10mm or so.

Draw a horizontal line across the centre of the flange, and make two punch marks at each end, between the outer and inner edges, so that you can tell which is the longer axis of the shape, later.

Making the External Tube

Cut the developed shape, but include a short extra straight section of perhaps 50 or 60mm at one end.

Using slip rolls, begin rolling the flat shape into a cylinder. When rolling, put the short extra section into the rolls first, as that will remain much flatter than the rest of the material passing through the rolls, because of the distance from the pinch

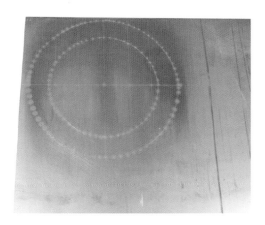

Fig. P4-5: Marking out for the flange.

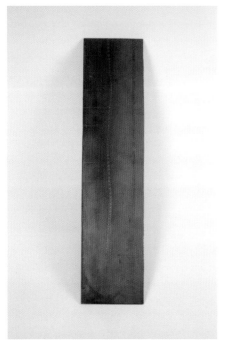

Fig. P4-6: Marking out for the external tube.

Fig. P4-7: Rolling the external tube.

Fig. P4-8: Checking the external tube for ovality.

point between the front rolls and the rear roll where curving begins. Make sure the scriber and punch marks are face down, so they will end up on the outside of the tube.

Gradually increase the height of the rear roll, and perform three or four passes at each setting (Fig. P4-7).

Using the long straight edge as a reference, continually check that the material is passing through the rolls at right angles, to prevent it twisting and becoming conical. When the strip is almost closed, cut off the extra straight section, then continue rolling until the ends meet. Remove from the rolls and check the diameter.

The tube will be completed by welding, and the fit around the hose cuff at this stage should be snug. Grind a V along the edges where they meet, then tack weld closed. Check the fit once again, and if satisfactory, weld the tube closed. Dress the weld inside (perhaps using a die grinder) and out (using an abrasive disc in a hand-held grinder).

Use a pair of inside callipers to check for ovality (Fig. P4-8), and correct by balancing on a high point and tapping twice, with a hammer. Check again. Mild steel is relatively soft, so corrections do not require great force.

Take care with this stage, as the rest of the construction is much easier if the tube is true and the hose cuff fits snugly (Fig. P4-9).

Fig. P4-9: Test fitting the hose cuff into the spigot.

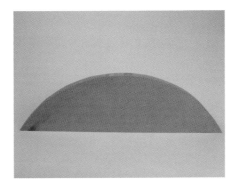

Fig. P4-11: Template.

Fig. P4-12: Completed external tube and flange.

Create the Flange

Use a hole saw to create a hole inside the inner section, then use a jigsaw to cut to the punch marks, or just inside. Use a coarse jigsaw blade and cut in an unhurried fashion.

Use the jigsaw to cut the outline, or just outside.

Use a grinder to dress the outline, but leave the inner hole undressed at this time.

Cut a template to match the outer curve of the vacuum cylinder, perhaps out of MDF or hardboard or plywood (Figs P4-10 and P4-11); or just use the actual cylinder as a reference, then use a set of rolls to bend the flange so that it has a horizontal curve to match the cylinder.

If the rolls will not bend the first section enough, be prepared to dress that using a hammer and a handy piece of pipe or the beak of a large anvil.

Weld the Flange to the External Tube

Take great care to orientate the flange correctly in relation to the tube, using the twin punch marks on the flange, and the scribed lines on the tube as a guide.

Use a die grinder or a file to dress the inner edge of the flange so that the tube slips snugly inside the flange. Check against the vacuum cylinder, as it is easy to misorientate the flange and tube at this point. Then double-check.

Tack weld in position (with four tacks), check and adjust as necessary. Then complete the welding, using stitch welds, in short sections opposite each other, until the whole weld is complete (Fig. P4-12). If the fit between tube and flange is good, the weld can be completed without adding much filler rod.

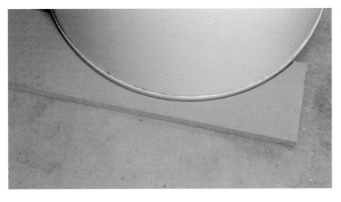

Fig. P4-10: Scribing a board to act as a template.

Dress the weld. Check that the completed assembly fits neatly against the vacuum cylinder. Spot through the flange bolt holes on the cylinder to mark the corresponding positions on the flange, then drill the holes.

Make the Inner Sleeve

There are two versions of the inner sleeve. The first version is relatively plain, and the key feature is that the outer diameter remains the same along the whole tube. The second version adds a wired edge at the rear, and a copper sleeve around the hole for the little plastic pipe. This second version can only be used if the hole in the vacuum cylinder wall is large enough to allow the wired edge to pass through. Otherwise, the plain version must be used.

Create the Basic Sleeve

The outer diameter of the inner sleeve should match the inner diameter of the external spigot. To calculate the developed width of the inner sleeve, use the inner diameter of the external spigot minus the thickness of the material used for the inner sleeve, so that the width is calculated using the neutral axis (nominally situated at the centre of the thickness of the sheet). For a cylinder of external diameter 104mm and wall thickness 0.7mm, the diameter at the neutral axis is 103.3mm. The developed length is π × 103.3 = 324.5mm.

To allow some adjustment and to avoid undue distortion, use a joggled flange. Note that this does not run the full length of the sleeve, and is omitted at the front where the burr will be formed, and for approximately 8mm at the rear to allow for the wired edge or the edging tape. Add one flange width to the developed width, for the joggled flange (the width depending on your jenny's rolls or the jaws of your flanger or joggler). That flange will also allow a little adjustment of the diameter of the sleeve, so you may wish to subtract 1mm from the developed width plus flange width, to allow adjustment inwards as well as outwards.

To the length, add 8mm or so for the stop flange at the front of the sleeve; for the wired-edge version only, add 2.5 × diameter of the wire to the rear.

Create the hole towards the rear, using a sheet metal hole saw, step drill or panel punch. Make its diameter large enough to take the pipe, 1–2mm clearance on the diameter for the pipe, and an allowance for the edging tape or the copper tubing around the inner edge. Although this hole will become elliptical in plan view after rolling, the 1–2mm clearance should take care of that.

Create the joggled flange, then roll the sleeve with the joggled edge initially towards the rear of the rolls. The flange may tend to flatten a little if it passes through the rolls, so it is best to wait until the sleeve has been rolled closed before finally rolling the flange and overlap together. Ease the pressure on the rolls at that point.

Place the sleeve inside the external spigot, then spot weld closed along the flanged joint so that it fits snugly.

Use the jenny with the burring dies to create the front flange. Note that this is folded inwards, so you may need to swap the positions of the burring dies to allow

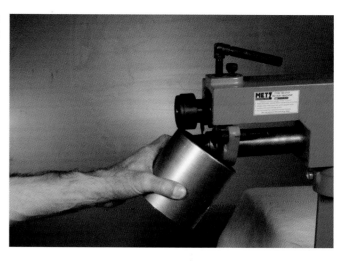

Fig. P4-13: Creating a burr (small flange) using the jenny.

Fig. P4-14: First stage in creating the flange on the end of the sleeve.

the sleeve to be folded downwards as the flange forms (Fig. P4-13). This is a tight fold, so go carefully but firmly. The sleeve walls may tend to spread at the joint, but can be persuaded back into the external spigot.

If the jenny's dies struggle to complete this flange (Fig. P4-14), complete it by hand using the fold line as a guide, and an upturned section of tube or bar as a dolly (Fig. P4-15).

Add an Optional Wired Edge

For the version with the wired edge, form a suitable length of wire into a circle, perhaps by using the grooves in your rolls. Then use the jenny to wire the rear edge with the wire on the outside. This is a tight squeeze because of the relatively small diameter of the sleeve.

Check beforehand that the securing nuts on your jenny's dies are recessed and will allow the die to run against the circumference of the sleeve without scuffing (Fig. P4-16). Roll a ring of wire, fit it into the

Fig. P4-15: Using an upturned tube as a dolly to finish the flange.

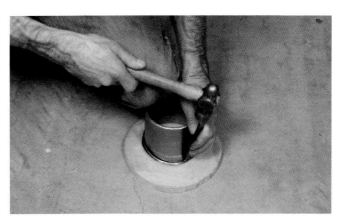

Fig. P4-17: Seating the wire loop in preparation for closing the wired edge.

Fig. P4-16: The nuts on the ends of the shafts of the jenny must not protrude beyond the ends of the dies.

groove, then close the groove over the wire using a suitable punch (Fig. P4-17).

Add a Safe Edge to the Pipe Hole

The pipe hole needs a safe inner edge, and on the plain-ended sleeve that can be achieved using folded panel edging produced for this kind of job (Fig. P4-18). Cut it to length, then slip it over the inner edge of the hole.

On the wired-edge version of the sleeve the safe edge can be produced from a slit ring formed from a piece of copper tubing, as shown in Fig. P4-19.

Cut a length of thin copper tube and bend it into a circle of a diameter suitable for the tube to sit tightly inside the pipe hole, with its inner face against the edge of the hole. Theoretically, the inner diameter of the ring will be the diameter of the pipe hole minus twice the wall thickness of the tube, but check that this will still allow enough clearance around the air pipe.

Slit the tube around the circumference, using a small hacksaw with a blade at least the thickness of the cylinder wall.

Using pliers, and working on one end of

Fig. P4-18: Panel edging strip.

Fig. P4-19: The cut edges of the hole are finished using a length of rolled copper tube slit horizontally around its circumference.

the tube, ease one corner outwards, then thread the tube on to the pipe hole. That ring can remain loose, or can be fixed in place by soldering (but leave any soft soldering until the sleeve has been fixed in place).

If the rear of the tube is to be fitted with edging tape, roll the bead near the end of the sleeve. Because the bead faces inwards,

Fig. P4-20: The finished spigot, flange and inner sleeve.

the dies in the jenny should be arranged so that the female die is inside the sleeve.

Position the sleeve inside the external spigot at the appropriate depth, and hard solder in position, or rivet the tubes together then put a little sealant around the joint at the rear. Before soldering, the mating surfaces must be thoroughly cleaned and fluxed prior to assembly.

Heat the outside of the external spigot and flange, rather than the inside sleeve, then apply the solder to the joint. Clean the joint after soldering. The finished piece should be a handsome replacement for the broken original (Fig. P4-20).

FINISHING OFF

Finish the job by painting the outer spigot and flange. Painting the inner sleeve is optional.

Project 5: Car Exhaust Expansion Chamber

Car exhausts come in different shapes and sizes, and they come in different types. Some exhausts incorporate an expansion chamber (Fig. P5-1) so that the hot gas can expand and lose some of its sound energy before leaving the exhaust. This makes the exhaust quieter.

The basic construction of an exhaust is quite simple, and they can be made from mild steel or, for a longer life, from stainless steel.

THE DESIGN

Fig. P5-2 shows a drawing of the expansion chamber disassembled. The gases pass down through a perforated tube and out of the end. They can't get directly through into the next little tube because of the blanking plate, so they have to go round the tube and enter via the little perforated holes. Then they can travel out of the end of the exhaust.

The area between the perforated tubes and the inside wall of the larger outer tube which forms the cylindrical body of the chamber is packed with Kevlar mat. Any sounds leaving the tubes via the perforated holes will be muffled by the Kevlar. That is why the entry into the second tube is blocked, so that the gases have to pass through the Kevlar to dampen the sound.

The perforated tubes are supported and held in place by the flanges, which are welded to the tubes and to the outer casing.

The whole exhaust is made of stainless steel, so it requires sharp tools and powerful punches to cut the tube and form the flanges.

Fig. P5-1: A typical exhaust expansion chamber designed to absorb sound.

Expansion chamber

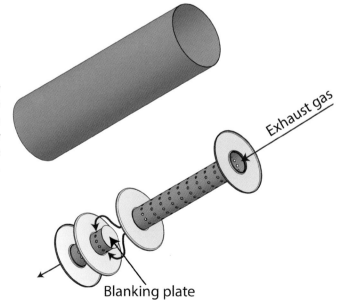

Fig. P5-2: As exhaust gases pass through the chamber, soundwaves are absorbed by the filler material outside the perforated tubes.

Project 5: Car Exhaust Expansion Chamber

Fig. P5-3: Using a cut-off saw.

Fig. P5-4: Chamber outer tube cut to length.

Fig. P5-5: Flanges formed and ready for welding to the tube.

To make welding easy, use a press tool to create a flange on the outside of the discs, and a second tool (*see* Figs 10-10 and 10-11) to press a generous entry radius around the holes in the inner flanges. Fig. P5-5 shows some finished flanges.

Welding the Inner Perforated Tubes

Insert the perforated pipe into the flanges and weld. They don't need to be fully welded all the way round every joint. Weld the small piece of sheet on to the end of the smaller pipe, to block it off (Fig. P5-6).

Assembling the Silencer

Slip the tubes into the silencer body and weld the outer edges of the flanges to the inside of the body tube (Figs P5-7 and P5-8). Then stuff the body full of chopped Kevlar mat (Fig. P5-9). – but make sure you use appropriate dust-collection equipment and a good mask.

Put the end caps on (noting that the flanges face outwards), and fully weld them to the rim of the body tube (Fig. P5-8).

The welds between the ends of the perforated tubes and the end caps will be completed as the silencer is fitted to the rest of the exhaust system.

MATERIALS

- Stainless steel tube of a diameter to suit the design (150mm in this project)
- Perforated stainless steel tube (50mm in this project)
- Small piece of stainless steel sheet 1.2mm thick × 50mm square
- Stainless steel sheet 2mm thick for flanges and end caps
- Kevlar mat

MAKING THE SILENCER

Preparing the Tube

Cut the tubes to length (Figs. P5-3 and 5-4), and dress the ends by deburring carefully.

Preparing the Flanges

Using a press and suitable tooling (*see* Fig. 5-17), blank out four flanges, then punch holes in the blanks to suit the diameter of the perforated pipe.

142 • Project 5: Car Exhaust Expansion Chamber

Fig. P5-6: The inner supporting flanges welded to the perforated tubes. The smaller tube has a blanking plate welded to the inner end.

Fig. P5-9: The chamber is packed with Kevlar mat. The internal section between the ends of the perforated tubes was packed before the shorter tube was welded in place.

Fig. P5-7: The longer of the perforated tubes is welded into one end of the outer tube so that its free end is flush with the end of the chamber.

Fig. P5-8: The flange on the shorter tube is welded inside the chamber so that the end of the tube is flush with the end of the chamber.

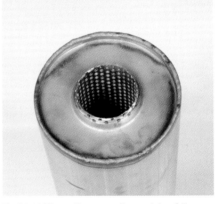

Fig. P5-10: The end flanges are inserted then fully welded to the outer tube. The inner welds will be made as the chamber is fitted to an exhaust system.

Project 6: Panels for a Sack Barrow

Fig. P6-1: The bare barrow.

Fig. P6-2: The barrow, refurbished and fitted with additional support panels.

The sack barrow shown here is in constant use, and it has had a hard life (Fig. P6-1). It's a handy device, but it has one failing: things often fall through the gaps between the bars, and the gap between the folding base and the upright back. The solution is to fill those gaps and provide more support for sacks and for small items.

THE DESIGN

Sack barrows differ in size and construction, but Fig. P6-1 shows the barrow used in this project. Fig. P6-2 shows the barrow all dressed up and ready to go.

The lower panel sits down on the small cross tubes. There is a tongue protruding at the front, and the rear of the panel is recessed at each side, to clear the hinges for folding the barrow flat. The lower panel is held in place by tabs that curl under the two main side tubes (Fig. P6-4).

The vertical back panel has two vertical flanges running down the sides. It curls around the top bar, and has supporting brackets that sit over the lowest tube, one at each side (see Fig. P6-13).

The back panel is locked in place by a sliding piece on the rear which is pushed against the underside of the top tube and prevents the panel from moving up or down (see Fig. P6-16).

MATERIALS

- Mild steel sheet 0.7mm
- M4 threaded rivet (one off)

MAKING THE PANELS

Marking Out

The barrow is made of tube and is intended to be a symmetrical shape. However, the base is a bit 'squint' at the back, so the measurements at the rear were slightly different on each side.

Using the dimensions from a sketch, the base was marked out directly on to the sheet using a ruler, square, dividers and scriber. Setting out for the bend AA' (Fig. P6-3) was straightforward. But bend BB' lies at an angle, so the dividers were used to create an arc at the front and rear ends of bend AA', equal to the estimated distance from line AA' to the outermost tip of the front or rear edge of the tabs; then a line was drawn tangent to both circles to find the front and rear corners of line BB'. The recesses between the tabs were aligned with reference to line BB'.

Once the panels had been marked out, a thick marker pen was used to place a line just outside the outline, to aid visibility and act as a guide to avoid mistakes when cutting out (Fig. P6-4).

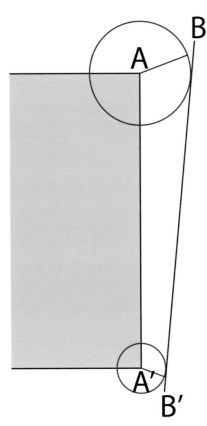

Fig. P6-3: Scribing circles to locate a tangent line.

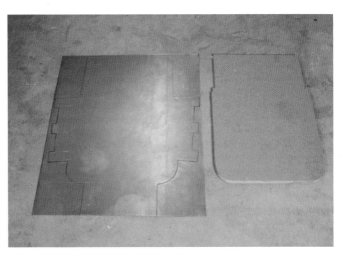

Fig. P6-4: Marked-out sheet, with additional pen lines just outside the scribed lines, as a safeguard.

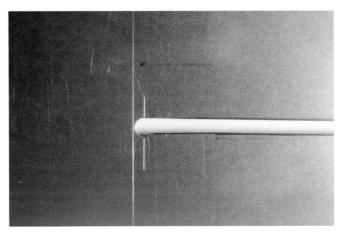

Fig. P6-5: Cut terminated by running out into a 5mm hole.

Cutting the Panels

Two rectangles, slightly larger than the base and rear panels respectively, were cut from sheet steel 0.7mm thick. Powered shears, snips and a hacksaw were then used to cut to the scribed lines. Several of the internal corners were formed by first drilling a 5mm hole, which was sufficiently large that the powered shears could finish a cut by running into the hole (Fig. P6-5).

The internal cut-outs in the sides of the base could have been produced using a small corner notching tool, but were instead created using snips, using the sequence of cuts as shown in Fig. P6-6.

Forming the Base

The base is an awkward shape because of the tapering sections on the sides. The

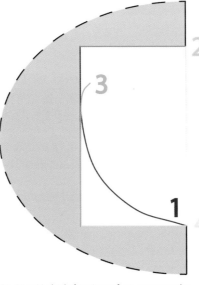

Fig. P6-6: Method of cutting to form a rectangular recess.

internal fold can be made on a straight or a box and pan folder, as can the internal fold at the front. The curved rolls over the tubular sides of the base of the barrow were made by first cutting an MDF panel to sit tightly in the base (Fig. P6-7).

Kneeling or standing on that panel to hold it firmly in place, the sides of the base sheet were formed over the side tubes using a strip of wood and a hammer, with final tuning carried out with a wooden mallet. At that stage, the reverse curve under the tubes was not formed, leaving the base freely removable. The two L-shaped locating stops were held in position and spot welded (Fig. P6-8). These act as stops to prevent the base sliding backwards or forwards.

Project 6: Panels for a Sack Barrow • 145

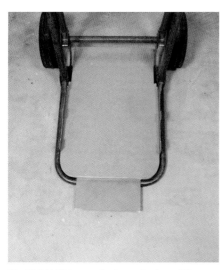

Fig. P6-7: Hold the wooden former in place while forming the sides of the base over the side tubes of the barrow.

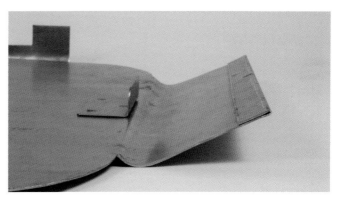

Fig. P6-8: Base locating stop welded in place.

Fig. P6-9: Begin creating a safe edge by folding the edge beyond 90 degrees.

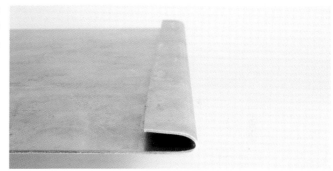

Fig. P6-10: Complete the safe edge by flattening the bent lip.

Fig. P6-11: Remove the rivets which hold the bottom of each leg in position, to allow the back panel to be fitted.

The safe edges at front and rear (Figs P6-9 and P6-10) were formed by folding to an acute angle, then closed by hammering on a wooden strip placed on top of the fold.

With the base in position, held by the former, the projecting front panel was bent over the front tube using a wooden strip and a hammer, until it touched the floor when the barrow was sitting on its base.

Forming the Back Panel

The two sides of the panel are simple right-angled folds and, unlike the sides of the base, are not formed around the tubes. Instead, they are created in a folder.

After forming the safe edge on the projecting bottom section of the panel, the front legs of the barrow need to be released by removing the rivets which hold them to the bracket at the bottom, to allow the panel to be put in place (Fig. P6-11). If these are hollow tubular rivets, use a large-diameter drill to remove one of the heads on each rivet, stopping as soon as the head comes off, and before causing damage to the bracket. During subsequent assembly and removal of the rear panel use temporary bolts to secure the legs.

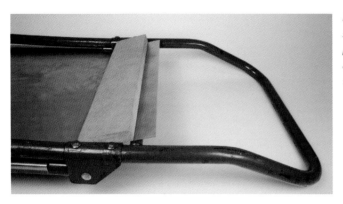

Fig. P6-12: Use a wooden strip to hold the top of the panel securely to allow the edge to be shaped over the top tube.

Sit the rear panel in place, mark the position of the upper fold, then form that in a folder. Then use one MDF strip to hold the upper part of the panel in place (Fig. P6-12), and use a hammer and a second MDF strip to form the curve over the top tube.

Mark the position of the bottom bend, and form that.

Hold the panel in position and fit the two lower locating brackets using spot welds. Support each bracket where it sits on the tube, and use a mallet to form the 90-degree curve over the tube (Fig. P6-13).

To prevent the rear panel from moving upwards and coming loose, the lock bracket should sit so that when the lower end of the box is almost touching the lower edge of the bracket, the top of the box is firmly against the underside of the top tube of the rear frame. When the box is removed, there should be sufficient space between the top of the bracket and the bottom of the tube to allow the panel to slide up and release both the top curve and the lower curved brackets, allowing the rear panel to be removed.

Position the box over the lock bracket, off the rear panel, and mark the position of the threaded pop rivet. Remove the lock bracket, drill the hole, then fit the rivet to the bracket (Fig. P6-15).

Mark the position of the bracket on the rear panel, and spot weld it to the panel.

Fig. P6-14: Lock box and bracket.

Fig. P6-15: Fitting the rivnut.

Fig. P6-13: Use a mallet to shape the lower locating brackets over the bottom tube.

Fig. P6-16: Lock box fitted, and preventing the rear panel from coming loose by moving upwards.

Making and Fitting the Locking Box

The panel should now sit securely on the rear tubular frame, but can still be removed by sliding it upwards.

Fold the lock box and the lock bracket (Fig. P6-14). Drill the hole in the lock box.

Check the lock box holds the rear panel firmly in position.

Painting and Assembling

The barrow had seen a hard life, so it was cleaned using a small grinder fitted with a 60-grit flap disc, dusted and degreased, then given three coats of a hard-wearing paint.

The rear panels were cleaned with steel wool and fine emery paper. The edges were rounded using a file and the 60-grit wheel on the grinder. The panels were dusted and degreased, then sprayed with a contrasting colour of the same brand of hard-wearing paint.

The base panel was fitted, then secured in position by forming the tabs around the side tubes, curling them under to grip the sides.

The rear panel was fitted along with the lock box.

Instead of fitting tubular rivets to secure the legs in the bottom brackets, two modified bolts were fitted. These had large heads on one side, and were threaded internally at the other end, to receive a small screw and a large washer (Fig. P6-17). This allowed for the possibility of removing the legs and the rear panel.

Extensive testing with bags of sand proved that the panels have transformed the performance of this heavily used and much-loved sack barrow. It looks even better than it would with factory-fitted parts, but it might benefit from a couple of tasteful go-faster stripes.

Fig. P6-17: Bolts at the bottom of the legs, to replace rivets removed earlier.

Project 7: Folding Steam Iron Shelf

A steam iron is an essential tool to maintain sartorial elegance, but it can be difficult to find a suitable place to store it in a small utility area. A folding shelf (Fig. P7-1) gives the iron a place, but can be folded out of the way (Fig. P7-2) when the iron is being used elsewhere. This is a light-duty shelf suitable for a small to medium-sized iron that is stored without containing water.

Fig. P7-1: Folding steam iron shelf in use.

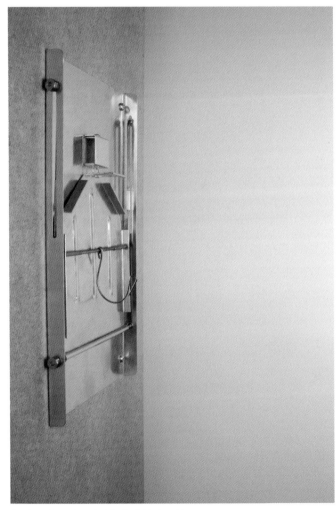

Fig. P7-2: Steam iron shelf in the folded position.

THE DESIGN

The details of the basic design (Figs P7-3 to P7-6) can easily be modified to suit the materials at hand, and the adjustment on the side-support arms means the sliding mechanism can still work even if other sizes are changed.

Fig P7-7 shows a mock-up of the shelf reduced to its basic components of shelf with hinge, side rails and side struts, and the spring catch to hold the shelf in the closed position. Note that final retaining collars for the hinge pin and arms are not shown in this photo, because the mock-up was made simply to test the operation of the struts and guide slots.

Provided that the physical relationships between hinge points, struts and slots are maintained, other details of the design may be varied to suit the iron or the location of the shelf. Figs P7-8, P7-9 and P7-10 show an alternative design of side strut, which is more complex to fold but has the advantage that the ends of the strut point inwards away from a user's hands.

Other arrangements, including the use of model aircraft push-rod ball-joint fittings, are possible. In the recommended design, the spring latch is incorporated within the back plate, but the separate spring box as shown in Fig. P7-11 works well in cases where the backing plate design has been changed.

MATERIALS

- 1mm aluminium sheet
- 3mm (⅛in) mild steel rod (*eg* welding rod)
- 1.6mm (¹⁄₁₆in) mild steel rod (*eg* welding rod)
- Brass tube, 3mm ID (or sized to provide a close sliding fit for the mild steel rod)
- M3 washers (brass, if possible)
- Collets (as used for model aircraft undercarriage), or fine soft copper wire and beads of 8 to 10mm diameter (option)
- Piano wire (0.6mm diameter)

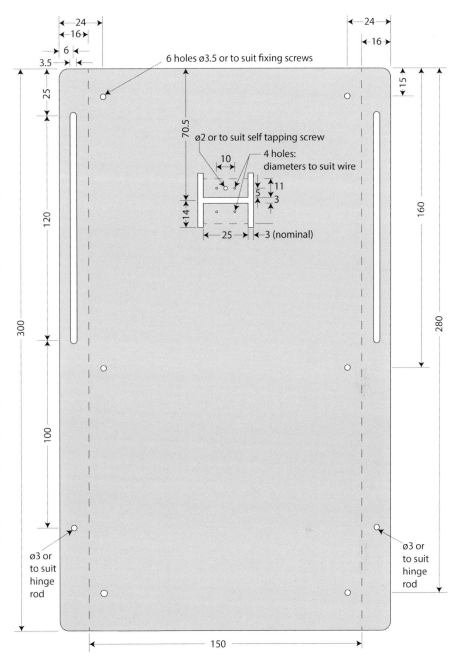

Fig. P7-3: Base plate flat pattern.

MAKING THE SHELF

Making the Vertical Base Plate

Using 1mm aluminium sheet, cut out the flat developed shape of the base plate, then deburr carefully. The top and bottom corners of the sheet, which, when folded, will become the corners of the side rails, should be rounded (for safety) or could be shaped artistically.

There is a choice of methods for holding the spring catch. The front and rear faces of the holder can be made integral with the

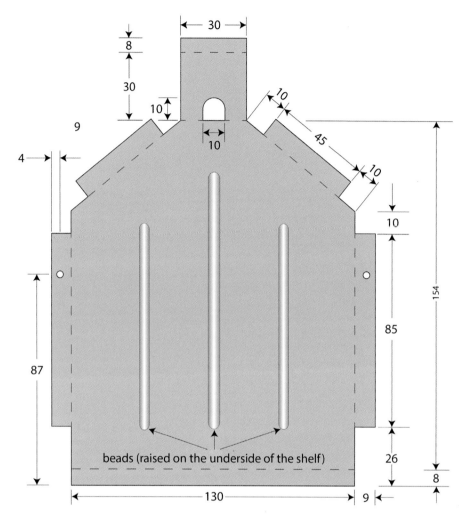

Fig. P7-4: Shelf flat pattern, showing optional beads.

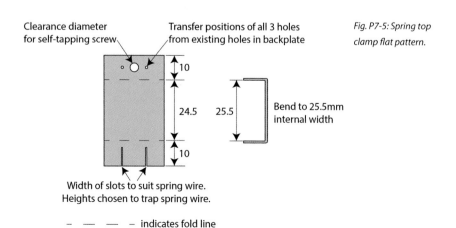

Fig. P7-5: Spring top clamp flat pattern.

base plate – though this method requires careful bending, and the work cannot be bent in a conventional box and pan folder. Alternatively, leave that section of the base plate flat, and make a separate spring holder, as shown in Figs P7-7, P7-8, P7-9 and P7-11.

Drill the holes for the shelf pivot rod and the mounting screws for attaching the plate to a vertical surface.

The slots for the support struts can be created by a variety of methods. Drill holes to create the ends of the slots, then use a fretsaw to remove the waste, and finish by using a thin flat file. Alternatively use an end mill or slot drill in a milling machine. A more challenging method is to chain-drill holes slightly smaller than the width of the slot, then saw and/or file between the holes. Finish the slot by filing.

If you have a fly punch with a thin rectangular punch, the main section between the rounded end holes of the slots can be created by setting a back guide and punching overlapping holes, finishing the ends by using a file.

If using the fly press, best results can be obtained by punching the slots before trimming the outside edges of the base plate to their final size, as the narrow strips left between the slots and the outside edges of the flat pattern are likely to distort unless they are firmly clamped.

Using a bend radius of 0.5–1mm, fold both side rails (Fig. P7-13) in the same way, by gripping the main sheet and folding the rails upwards, or by gripping each rail in turn and folding the base plate upwards. Either way, it is the consistency of the method that is important, so that the slots in the rails end up at the same height from the base plate.

The bend line indicated in Fig. P7-3 assumes gripping the base plate and folding the sides upwards, but the same bend line can be used for either method, because although the distance between the bent sides will differ slightly, the

Project 7: Folding Steam Iron Shelf • 151

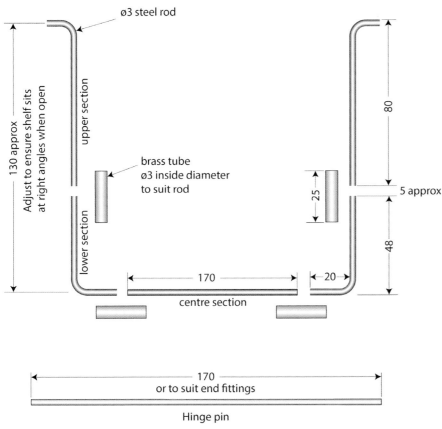

Fig. P7-6: Folding shelf struts and hinge rod dimensions.

is not critical, as long as their straight edges are a little more than the eventual height of each side.

The slots can be formed by drilling 3mm holes at each end, then using a fret saw (scroll saw) to join pairs of holes by punching or by milling. They could also be formed by chain drilling, then filing the slots to finished size using a flat needle file.

Drill the four small holes for the spring wire and the single pilot hole for the self-tapping screw. These holes are sized to suit the wire and are rather small, so some care is needed when drilling, to avoid breaking the drill bit. A tight fit between wire and hole is not absolutely essential, and they can be a little larger in diameter as the clip can be made to trap the wire in the bottom of the holes.

Note that it is difficult to finish the holes once the sides of the spring holder have been bent upwards, so make sure the holes have been fully formed before bending.

Prepare for bending by marking out the bend lines on the reverse side of each tab. The dimensions given in Fig. P7-3 include a bending allowance for 1mm soft aluminium sheet, using a 0.5mm radius bend.

Open a vice to the distance between the bend lines, and align the bend lines with the edges of the vice jaws. The following method assumes a radius on the top edge

hinge arrangement for the shelf provides sufficient adjustment to cope with this.

To fold an integral spring catch, mark out, then cut the slots which define the edges of the two faces of the spring holder. The width of the slots is not important, as long as the inside edges are the correct distance apart, because these define the edges of the tabs, which will be bent upwards to form the holder. Similarly, the length of the slots

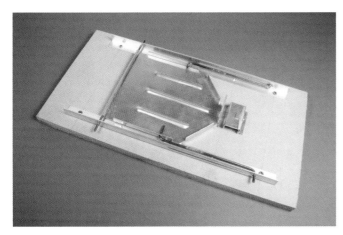

Fig. P7-7: Basic essentials of the mechanics of the shelf unit, showing shelf, side guides, struts and spring catch.

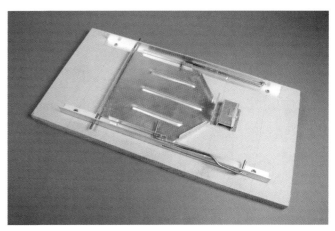

Fig. P7-8: An alternative design of side strut in place.

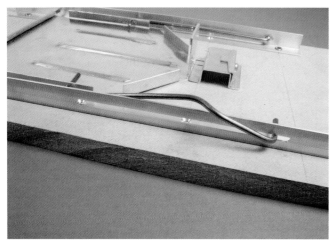

Fig. P7-9: The alternative side strut rises up over the side guide, and both ends face inwards.

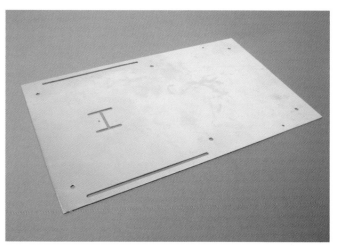

Fig. P7-12: Base plate ready for bending.

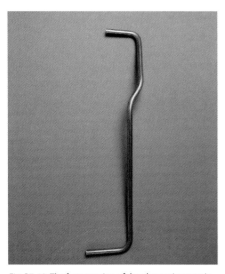

Fig. P7-10: The front portion of the alternative strut is also offset to the side.

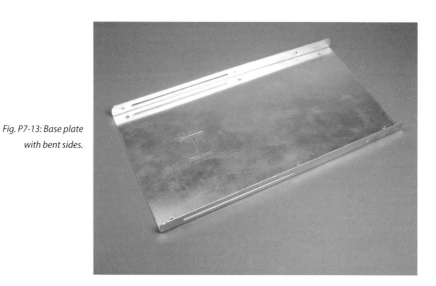

Fig. P7-13: Base plate with bent sides.

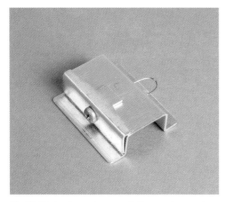

Fig. P7-11: An alternative, stand-alone, spring catch box.

Fig. P7-14: Wooden punch to bend the sides of the spring holder.

of the vice jaws equivalent to approximately 0.5mm. The bends can be made using accurate blows from a small hammer on to a simple wooden punch which has a width of 25mm to match the faces and a breadth of 23mm to match the distance between the insides of the faces once bent (Figs P7-15 and P7-16).

Although not essential, Fig. P7-16 shows that the wooden punch can incorporate a small raised ridge to fit into the slot cut between the faces before bending: this helps to locate the punch for the first few

Project 7: Folding Steam Iron Shelf • 153

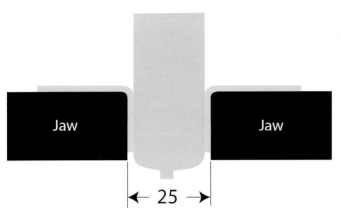

Fig. P7-15: The punch folds the sides so that the outer faces are 25mm apart.

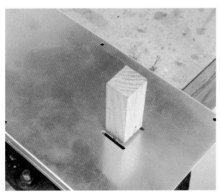

Fig. P7-16: Wooden block in position ready to make the two bends for the spring catch simultaneously.

Fig. P7-17: The bent sides for the spring catch.

blows, and helps ensure an accurate start to the bend. Drive the punch right down until its parallel sides force the faces of the spring holder into their final positions (Fig. P7-17).

Fig. P7-18: The completed catch, with the wire loop positioned to just catch the shelf.

Fig. P7-19: The top clip is bent to accommodate the width of the slot plus the diameter of the wire, with approximately 0.1mm interference fit.

Fig. P7-20: Arrange the lengths of the sides of the top clip to avoid the internal and external radii as shown.

Make the spring holder top clip to suit the upstanding faces: it's a simple U shape (Fig. P7-18). Make the distance between the inside faces of the legs of the U the same as the distance between the upstanding faces of the spring holder, plus the thickness of the wire, to accommodate the bent tails of the wire clip which will eventually be trapped between the rear faces of the holder and the top clip, leaving an approximately 0.1mm interference fit (Fig. P7-19).

Make the top clip a little broader than the length of the upstanding faces, so that it overlaps and hides the edges of the faces. Make the downward faces shorter than the upstanding faces on the base, to avoid the bend between the bottom of the spring holder faces and the base plate, and on the inside of the top clip, between the horizontal and vertical faces (Fig. 9-20).

To transfer the positions of the spring clip holes from the upstanding faces on the base plate to the top clip, set a height gauge or a pair of callipers to the height of the holes above the base plate (Fig. P7-21). Place the top clip in position, and mark off at that height (Fig. P7-22). Drill the clearance hole for the self-tapping screw in the rear face, but do not drill holes for the spring clip itself, just yet.

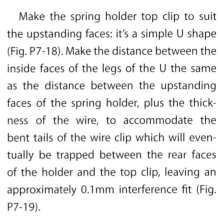

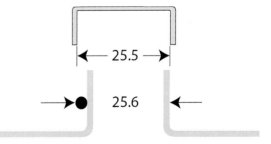

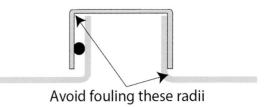

Avoid fouling these radii

Fig. P7-21: Using a height gauge to find the distance from the base plate to the spring holes.

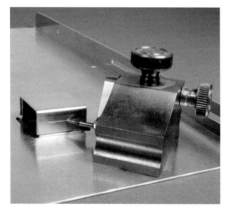

Fig. P7-22: Transferring the distance to the top clip.

The wire spring for the catch is best made once the shelf has been fitted, to ensure that it works effectively. The wire loop should protrude through the hole cut in the shelf where the handle is formed, and should catch just enough to require light operating pressure.

The spring is made by bending piano wire, inserting it into the holes, adjusting the distance that the front of the loop protrudes in relation to the front edge of the shelf so that the catch operates effectively, then bending the rear wire tails outwards. The top clip traps the tails, as well as sitting across the tops of the flanges, to finish the catch nicely.

At the front, use a piercing saw to cut two slots to accommodate the wire (see Fig. P7-5), stopping short of the centre line of the wire (marked previously, using a height gauge or callipers). When fitted, the slots in the top clip will straddle the wire. Adjust the tops of the slots so that the clip traps the wire without distorting it appreciably. This compensates for any difference between the diameter of the wire and the diameter of the drilled holes.

Making the Hinged Shelf

Cut the developed shape from 1mm aluminium sheet. Begin making the cut-out at the base of the handle by punching or drilling a pilot hole, then complete the shape by using a fretsaw; finish with files. The exact shape of this cut-out is not important, so the hole could be punched in a fly press using a rectangular punch, or a circular punch followed by a rectangular or square punch.

Drill the side holes for the hinge. The ridges provide additional strength for the plate, and can be formed using a jenny, but take care to leave enough plain sheet at the ends to accommodate the bending beam of the folder for making the subsequent bends without crushing the grooves.

Using a box and pan folder, form the safe edge for the handle, and make the vertical bend for it; then fold the sides by gripping each side and folding the main section of the shelf upwards, or by gripping the main sheet and folding the sides upwards. As when bending the side rails of the base plate, consistency of method is the key to accuracy here.

It is important that the vertical heights of the folded sides are no shorter than shown, as they are sized to provide enough depth for an M3 washer.

The wired edge at the rear of the shelf acts as a hinge, so make this a good fit for the wire (Fig. P7-23). The fold line assumes 1mm aluminium and a 3mm (⅛in) diameter wire. The wire needs to be able to move to the right and left to allow assembly through the holes in the sides of the base plate.

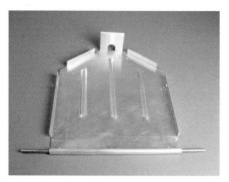

Fig. P7-23: The shelf with bent edges, wired edge for the hinge rod, and optional beads protruding from the underside.

The lengths of the spacing tubes at the sides of the hinge should be adjusted to allow the shelf to sit centrally between the sides of the base plate, allowing perhaps 1mm clearance across the whole hinge (0.5mm each side), or less. The shelf should be able to swing freely, and the side washers are intended to assist this motion and prevent the edges of the tubes jamming against the sides of the base.

Making the Side Struts

The critical feature of the side struts is that when the shelf is at right angles to the base, each strut is at the bottom of its guide slot. This condition should apply when the shelf is fully loaded, so the iron is never sitting on a shelf that slopes downwards at the rear. Both struts need to be fully seated at the same time, so that the shelf is held level (side to side).

The struts could be accurately bent in a jig, but for a single shelf unit, the operating lengths of the struts can be adjusted using the tube sleeve, so that the position of the shelf can be optimized easily.

From an aesthetic point of view, the length of the sleeve can be varied, as can its position; one possible variation might be to make the tube span almost the full length of the strut. The end rods only need to protrude 10mm or so inside the tube.

Bend the ends of the arms using a simple

Fig. P7-24: The ends of the wire struts can be bent in a simple multi-purpose jig.

Fig. P7-26: One method of securing the hinge rod is to use collets, as supplied for model aircraft undercarriages.

jig (Fig. P7-24), judging the right angle by eye, then by using an engineer's square. Adjust as necessary.

Using soft solder, tin the ends of the arms where they will sit inside the tubes, then dress lightly with a file so that they slip inside the tubes. Cleanliness is essential when preparing to tin, as is the use of a flux. Flux the insides of the tubes, and solder one end of each strut into its tube.

Set the shelf in its operating position, but in the absence of the iron, make the angle between shelf and base a little less than 90 degrees. Put a strut in position and adjust the unsoldered end so that it is right at the bottom of the guide slot. Mark and remove it, then solder it in position. Check the operation of that strut, then repeat the sequence for the other strut.

Check that the shelf is sitting level, and adjust the struts as necessary.

Add the washers to the struts. Brass washers are recommended, as they take solder much more effectively than plated washers, although plated washers can be used, with care.

Fig. P7-25 shows a simple jig that ensures the faces of the washers sit at right angles to the axis of the struts. It is simply a hole drilled in a piece of scrap material, preferably material to which solder will not stick easily (such as aluminium).

The ends of the struts can be finished in a variety of ways. The object is to secure the struts so that they remain in the guide slots and resist a small amount of side pressure as they operate; then to finish the ends of the rods neatly and safely, leaving no sharp edges to catch fingers or fabric. The struts on the shelf unit shown in Fig. P7-1 were finished by trimming the protruding ends to length, and securing them using collets of the type fitted to model aircraft undercarriages (and often used for other jobs).

There is a washer between the inner face of the collet and the side of the base, to ensure an easy operating action.

Alternative finishing approaches include soldering the washer in position, then drilling and gluing a bead (from a craft shop) on to the rod to cover the end. The washer takes the side thrust, and the bead adds decoration while covering the rod end.

Making the Hook for the Power Cord

The hook (*see* Figs P7-1, P7-2 and P7-27) accommodates the power cord for the iron and is made using 1.6mm rod bent to shape using pliers, and by forming round rod of a suitable diameter. The diameter of rod is not critical, nor is the shape, provided the finished hook can swing freely and accommodate the power cord for the iron. There should be no protruding ends or sharp edges in the finished hook.

Figs P7-27 and P7-28 show that the hinged end is formed so that it wraps twice around the hinge rod once assembled; this ensures that it will not unwrap under load. The shape of the hook determines how it hangs, so some experimentation may be required to ensure it hangs in a convenient position to be able to quickly put the cord over the hook.

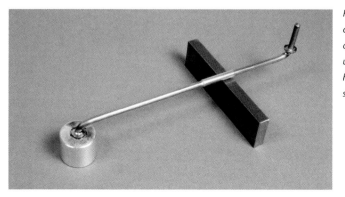

Fig. P7-25: The washers can be soldered in the correct orientation by using a jig consisting of a hole drilled in a piece of scrap aluminium.

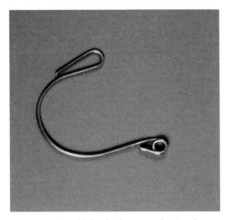

Fig. P7-27: The cable hook is bent to a pleasing shape, and the cut ends are tucked out of the way.

Fig. P7-28: The completed strut and rod assembly showing the hook in position.

Assembling and Fitting the Shelf

Assembly of the shelf is best carried out on a flat surface.

Prepare the tubes and the central rod to join the lower ends of the struts. Note that there is a washer at each end, between the faces of the tubes and the inner sides of the shelf.

Flux and tin the ends of the centre rod. Slip the rod inside one tube, and fit the assembly over one strut end. Slip the rod Inside the tube far enough to be able to assemble the hook and the other tube, before sliding the tube back outwards over the end of the other strut. Don't forget the washers or the small amount of clearance at each end.

With everything in position, apply heat to each of the four joints in turn.

FINISHING OFF

Deburr all edges, especially round any sharp corners, and clean any flux or solder residue.

If the sheet has become marked, treat any deep scores using wet and dry paper in progressively finer grades, followed by crocus paper or micromesh. Finish with a polishing cream such as Solvol Autosol.

Mount the shelf unit in its final position using appropriate screws. Put the iron on the shelf, then stand back and admire your work. It would be a pity to have to take the iron off the shelf and actually do the ironing.

Further Information

ARMOUR

Age of armour
www.ageofarmour.com

Corrugated iron artworks
www.corrugatedcreations.co.nz

FORUMS

The Sheet Metal Shop
www.thesheetmetalshop.com

Metal Meet
www.metalmeet.com

All Metal Shaping
allmetalshaping.com

The Home Machinist
Sheet metal fabrication:
www.chaski.org/homemachinist

Metal Artist
www.metalartistforum.com

MACHINERY AND TOOLS

RMT-Gabro and the M J Allen Group of Companies
www.mjallen.co.uk

EDMA Outillage
www.edma.fr

TRUMPF Group
www.trumpf.com and www.uk.trumpf.com

Warren Machine Tools Ltd
www.warco.co.uk

RIDGID Tool UK
www.ridgid.co.uk

Irwin Tools
www.irwin.co.uk

JD Squared Inc.
www.jdsquared.com

Jack Sealey Ltd
www.sealey.co.uk

Machine Shop Services

John Saunders at NYC CNC
www.NYCCNC.com and on Facebook

SHEET METAL MERCHANTS

Smith Metals
www.smithmetal.com

Aalco
www.aalco.co.uk

Ayrshire Steels
www.ayrshiresteels.co.uk

STAINLESS STEEL EXHAUSTS AND CUSTOM FABRICATION

RMS Engineering Ltd
www.rmsexhausts.com

TECHNICAL DRAWING (MANUAL, NON-CAD)

Pickup, F., and Parker, M. A. (1960) *Engineering Drawing 1*, London, Hutchinson Technical Education

Dickason, A. (1967) *The Geometry of Sheet Metal Work*, London, Pitman Publishing

TINPLATE

Tinplate Girl: information, instruction, plans and projects for tinplate work
www.tinplategirl.com

Index

2D 23
3D 23

abrasive 110, 115
acid 111
alloy 11
aluminium 11
American Wire Gauge (AWG) 13
annealing 101
anodising 15
assemblies 23

bandsaw 43
bead roller (*see* jenny) 77
beading 77
belt sander 48
bench peg (bench pin) 46
bend 26
    allowance 29, 69
    angle 65
    deduction 69, 70
    line 61, 150
    minimum bending factor 66
    minimum inner radius 66
    over-bend 65
    radius 26, 62, 150
    relief 26
    size 61
Birmingham Gauge (UK) 13
blanking 56, 57
    plate 140
blasting (bead, sand, shot) 111
bolster 55
brazing 96
Brinell hardness 15
bronze 10
Brown & Sharp (B&S) 13

callipers 136
cartridge brass 16
centre 34
    centre mark 34
    centre punch 34
chrome 17
clamp 99
Cleco fastener 99
clips 86, 87
    U clip 86
coating
    dip 112
    powder 113
    metal 113
cold rolled 13
compasses 32, 33
computer-aided design (CAD) 23, 61
computer-aided manufacture (CAM) 29
computer numerical control (CNC) 29, 57, 59
copper 10
corner 26, 29
    notching 56, 57
    relief 26
corrosion resistance 15
crimping 78
cutting 29

deburring 46, 47, 48, 156
development 24, 135
die 54, 103, 104
dimension 24
dip coating 112
dividers 32, 33
drawing (deep pressing) 104
drawing (technical) 21
    standards 21
    projection 21, 22
    computer-based 21, 23
    views 21
    elevation 22
    plan 22
drill 51
    butterfly 53, 127
    conical 54
    geometry 51
    parabolic 52
    rake angle 52
    speed 52
    step 53
DXF (drawing exchange format) 25

elastic limit 61, 65
elasticity 61, 65
electrolyte 113
electroplate 18
elongation 15
EN specification 13
European steel specification 13, 14

faces 23
fastenings 83
feature 24, 27
ferrous 13
file 46, 150
fixings 83
flange 24, 78, 137, 140
    lofted 27
flanging 78, 106
flat pattern 27, 61, 68
fluidising 112, 131
foil 10, 13
fold 94, 128
    double grooved seam 94

Index • 159

fold together 94
grooved seam 94
interlocked edges 94
U fold 94
folder 62
 angle stop 67
 backset 63, 65
 backstop 67
 box and pan 63, 154
 cam clamping 64
 clamp beam 62
 finger 63
 gauge 64
 nose radius 65
 repetition aids 67
 segments 63
 side guide 67
 simple 62
folding 29
formability 17
former 101
fretsaw 46, 150

gauge 13
gold 10
grinder 110
guillotine 41

hacksaw 40
hardening and tempering 120
hole saw 53, 132
hot rolled 13

iron 10

jenny 77, 138, 154
jig 155
jigsaw 44
joggling 78, 93, 132
joint 94 (see fold)
 glued 98

K factor 68
 calculating from a test bend 70

laser cutter 48
lead 10
leaded brass 16

leaf 13
lock former 94
lofted flange tool 27

machinability 15
malleable 17
Manufacturers' Standard Gauge (USA) 13
marking-out 30
 fluid 30
mercury 10
milling 29, 59
molybdenum 17
Monodex cutter 38

naval brass 16
neutral axis 68
nibbler 43
nickel 17
nickel silver 19
non-ferrous 13
notcher 41
 Gabro notcher 41
 corner 42
nut 86
 caged 86
 captive 86
 rectangular flat 86
 U nut 86

pad saw 40
painting 111, 130, 139, 147
parts 23
piercing saw 45, 154
planishing 102
plasma cutter 48, 49
plate 13
plating 18
polishing 114
press 55
 arbor 55
 between guides 76
 brake 70
 fly 56
 hydraulic 55
 punch 34, 54, 57, 70
 C frame 56
 centre punch 34
 hydraulic 58

pricker punch 35
Q.Max 58, 129
Redman 56
press brake 70
 air bending 72
 bottoming 72
 coining 73
 die 70
 multi bending 73
 punch 70
 U bending 73
 V bending 72
primer 112
punch 10, 103, 104, 120, 133, 136, 152

rip tool 27
rivet 88
 blind 91
 dolly 89
 flat punch 89
 hole formula 89
 nut 91, 129, 146
 pliers 91
 pneumatic closing tool 92
 pop 91
 punch 89
 Rivnut 91, 129, 146
 sett 89
 snap 89
 squeezer 90
 stake 89
 tongs 91
rollforming 77
rolling 13, 75, 135
 combination bending roller 76
 slip rolls 75, 136
rope 19
ruler 30

safe edge 145
scriber 30, 32, 136
scroll saw 46
shrinking 10, 29, 81, 102
shearing 10
shear strength 15, 37
shears 37
 powered 39
 bench-mounted 41

sheet 13
Shetack saw 40
silver 10
sketch 23
skinpin 99
slip joint 79
snips 37
   aviation 38, 126
   compound action 38
soldering 94, 122, 139, 155
   acid 96
   flux 95
   hard 95
   iron 95
   silver solder 95
   soft solder 95
   tinning 95
soleplate 39
spring-back 65
   factor (SBF) 66
square 31
   combination 31
   engineer's 31
   framing 31
   roofer's 31

standard wire gauge (SWG) 13
stainless steel 17
steel 11
straight-edge 30
strain 61
stress 37, 61
stretching 10, 29, 81
striker 80
stripper plate 55
surface finish 15
swage (*see* jenny) 77

tap 84
   geometric 85
   names 84
   spiral 85
technical drawing 10
temper 15, 17
tensile strength 15
thickness 24
thread 83
   engagement 83
   pitch 83
tin 10
tinplate 18, 118

transformer 27
trepanning 58
tube 119, 134, 139, 140, 151
   bending 121
   perforated 140

washer 57
water jet cutter 49
welding 96, 137
   electric arc 96
   MIG 97
   oxy-acetylene 96
   qualities 15
   spot 96
   TIG 97
   torch 96
wire 19, 154
   piano wire 19
   nichrome wire 19
wiring an edge 79

yield strength 15

zinc 11
Zintec 13